A FISHERY OF YOUR OWN

A FISHERY OF YOUR OWN

Its Management and Fishing

Barrie Rickards and
Ken Whitehead

With 4 colour plates
82 photographs
and 27 diagrams

Adam and Charles Black
LONDON

First published 1984
A. & C. Black (Publishers) Ltd
35 Bedford Row, London WC1R 4JH

Rickards, Barrie
A Fishery of your own.
1. Fishery management—Great Britain
I. Title   II. Whitehead, Ken, *1930 Aug. 10–*
639'.21   SH255

ISBN 0-7136-2413-2

Printed in Great Britain by
R.J. Acford, Chichester, Sussex.

CONTENTS

ACKNOWLEDGEMENTS

Ken acknowledges the assistance he has received from Mrs J. R. Frost, hydrologist; Mike Harcup, Fisheries and Recreation Officer; and the bailiffs of the Welsh Water Authority. Equally helpful have been Dr Buckley, Sussex Fisheries Officer and Michael Clark, bailiff, both of the Southern Water Authority.

Barrie would like to thank his many angling friends who have been involved in management problems, small and large, over the years, and especially the officers of the Anglian Water Authority for much help and advice.

Our mutual thanks are due to Dr S. M. Haslam, who very kindly helped us with our plant identification—though any errors in this respect are entirely of our own making—and to an old helpmate, Ron Coleby, who not only deals in secondhand books gave us every assistance in compiling our bibliography.

vii

DEDICATION

*We dedicate this book to the growing army of
anglers whose interests and energies are devoted to
ecology, and all attendant branches of natural
history.
Their efforts will ensure that coming generations
will have considerably more than fishing to study and
enjoy, despite efforts from those with anti-Field
Sport beliefs.*

INTRODUCTION

Smallwater men are a breed on their own. They are usually restless characters who find the sitting and waiting in angling completely beyond their capabilities and who like to work for their fish. The more difficult those fish are to catch and the harder the conditions that may have to be faced before they can be reached, then the greater the mental stimulation and physical challenge.

Of course, fishing – the actual physical work with rod and line – is not the be-all and end-all of the sport. Most anglers make natural conservationists and for many the ability to manage a small water, to improve its stock and the surroundings whilst learning and absorbing the many facets of plant, animal and insect life soon becomes as absorbing and interesting as a day spent in actually landing fish.

Expense is another reason for fishing and managing the small water. Just how high that price can be is well demonstrated by adding together the *real* cost of a day's reservoir trout fishing on one of the better-known waters. Including permit, boat hire and travelling (say around 100 miles – not an unreasonable mileage) the angler is not likely to get much change out of £20, which lifts the value of the fish limit (universally in single figures) into the gold-plate bracket when it is compared with the price of trout on the fishmonger's slab.

Small waters are not greatly expensive to maintain; if run purely for coarse fishing then the outlay is likely to be negligible. Should the water be suitable and angler inclined to make it into a trout fishery, then the outlay still need not be prohibitive providing time and care are spent in getting the water up to scratch before fish are introduced – and where two or more anglers band together into a syndicate, then one can confidently predict that good sport will be enjoyed for a lot less in both time and expense than that spent by those who stick to fishing large waters.

Having outlined a few of the pleasures and some of the challenges to be found in managing and fishing a small water, what and where are its perils and pitfalls? The short answer is that both usually occur when the managers of a water stick to the hard and fast rules that would seem to have been handed on from the past. Because a pH value appears insufficient to support life, or weed is lacking or too dense, or the insect count proves practically non-existent it does not mean, as many books and articles suggest, that the water is incapable of being moulded into a fishery of one sort or another.

We both know of ice-cold and acid mountain waters where there is no weed, but an abundance of insect life, and of lowland waters where the pH value should, in theory, fail to support life of any sort, but where the weed is lush and flourishing. We could quote many similar examples but feel it could serve no useful purpose. The only way to manage a water is to have a go at the physical tasks as well as to think, reason and plan each attempt to introduce life into it.

Log your work in diary form and complement it with photographs so that the perils and pitfalls you may encounter are quickly recognised, and so avoided. Read everything you can lay your hands on that embraces the subject, and do not be afraid to adapt ideas which you may see. Whilst no two small waters are alike, every one of them is capable of improvement and of providing sport with some thoughtful and imaginative management.

A great deal of our time has been spent fishing small waters of all kinds, over much of the U.K. and elsewhere, and not much less in managing them in some capacity or other. In this volume we make no serious attempt to separate the fishing from the managerial work, as the two are often inextricably tangled – in preparing a swim immediately prior to fishing, for example. So you'll find bits of fishing in amongst the managing, where it's directly relevant, and *vice versa*. We begin in Part I with fathoming small waters, the fundamental geology, chemistry, ecology, and particularly the role of weed. Part II deals exclusively with the problems of finding a water, securing it, and beginning to manage it as a fishery. Part III considers fishing in small waters of different types, together with the kind of management problems that each type presents to the angler. Part IV is involved entirely with fishing, but with the fundamentals of tackle, coarse and game fishing as far as they are relevant to small water fishing. Finally, Chapter 17 and the list of books for further reference are designed to help the managing angler take things a little further. So the volume taken as a whole proceeds from the fundamentals of the water itself, through management practice, and ends with the fundamentals of fishing – with a fair degree of overlap throughout. We hope you enjoy it, find it valuable, and we leave you with two thoughts: fishing small waters is an art and a craft, not a science; and water management is so strongly empirical that it too is more of an art than a science.

Barrie Rickards
Milton
Nr. Cambridge

July 1983

Ken Whitehead
Herstmonceux
Sussex

PART I

FATHOMING SMALL WATERS

CHAPTER 1

GEOGRAPHICAL AND GEOLOGICAL SETTING

One of the most significant factors concerning managing and fishing small waters is that they occur almost everywhere in the country. It is unusual to find a village without its nearby pond – whether it is the result of a wartime plane crash or the surreptitious extraction of gravel or clay by the local farmer. Anglers, as we point out, are turning to small waters more and more in an attempt to find good, quiet fishing in less crowded circumstances, so it is as well to know not only *where* these waters are to be found, but why they are there. Knowing *why* may enable you to find or create others, and such knowledge may be critical in fishery management matters, particularly where water flow and chemicals are concerned.

It is unusual to find large tracts of these islands without small waters. Some highland areas – the Yorkshire Dales, for example – have relatively few ponds but do have many hill streams. The west of these islands probably has fewer ponds, when compared to the south and east England lowlands, but they have more streams. The Midlands have fewer ponds than the east, but more canal 'cuts'. The Lincolnshire and East Anglian fens have areas full of small gravel pits, yet large tracts with few small still waters – but with ample compensation in the form of miles of tiny drains, and the same is true of Somerset. In Scotland, and in the Lake District, there are areas packed with small lochs and tarns, whilst in other areas, particularly those of extensive higher ground (say over 1500 ft) still waters are less common and hill trout streams replace them in frequency.

In short, small waters of varied kinds are almost everywhere. Because they have been neglected by anglers the fishing is often virgin, and fishing permission can commonly be obtained with a minimum of formality. It follows, too, that if you are seriously interested in *acquiring* a water (see Chapter 4) you stand a much better chance here than with the large, up-market waters.

An important question is, how did the U.K. come to have these numerous small waters? Well, in the first place we have a wet island,

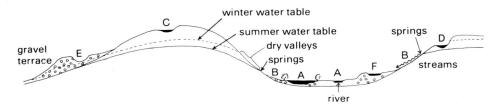

Fig. 1   The geological structure of a typical small water. (coarse circles = gravel; black areas = river silt/alluvium).

and despite bad drainage policies which waste most of the water (see Chapter 11) some of it sticks! Essentially, in order to form a small still water the hole must have its base below the water table, or its bottom must be impervious. We show this in Figure 1: streams and ponds can exist at A and B because the ground and rock below is permanently saturated with water, whereas at C the lake must have an impervious bottom, such as a layer of clay. Perhaps we should point out the manner in which water is held in rocks: it does not usually form underground rivers (except in parts of Yorkshire and Derbyshire, where the water opens out joints and cracks in the rock to form tunnels and caverns) but is more commonly held in similar fashion to water in a sponge. If you take a large sponge and make it damp, then place it on the table, the water quickly runs to the lower half of the sponge and begins to seep out on to the table. On the other hand if the sponge is really soaked, any water you add to the top will run off, not soak in. It is exactly the same with the countryside: rain that falls either soaks in (to become part of the water table as in Figure 1) or it runs off in the form of streams, or collects in ponds and hollows. (Some, of course, evaporates.)

Obviously, therefore, it may not be enough to dig a hole to obtain a lake. You *must* dig so that the base of the pond is well below the dry weather water table. In Figure 1 we show a high level gravel terrace on a valley side (site E). Any gravel workings here would invariably be dry because the base of the hole is above even the level of the wet weather water table. On the other hand, a gravel working at F, a low level gravel terrace, would almost always have a good supply of water. Quite a lot of small waters fished by anglers come into the category D of Figure 1, that is they are wet in the wet season (winter) but may dry out in the summer as the water table is lowered – *or as abstraction takes place*. The apparent permanency of a water viewed in the winter months is no guide at all: make enquiries as to the summer levels, especially from the local Water Authority or Water Boards.

Nature has many ways of forming lakes and ponds, and man has given them many more names, but the above basic principles apply: there must be a source of water (rain, or water table) and the water must be held (by seepage from the water table, or by an impervious layer such as clay or its baked, metamorphosed form, namely slate). Man-made kinds of C (Figure 1) are the *dewponds* in chalk country, as in Kent, East Anglia and Yorkshire: they have a lining of cement or clay and collect rainwater and dew. Some of them may be natural hollows lined with clay, and many of them have a good stock of small carp and tench. If the bottom of a pond is clay then the fishery manager need have no worry about nutrients for the water: plant life will flourish and if the pond is on chalk then a high pH is likely anyway.

In regions of gravel pits along valley bottoms or on low level gravel terraces (Figure 1, sites A and F) the water is assured and the chemistry dependent on the water which flows through the gravel layers: if the region has soft water then that in the lake or stream will be soft and liming may be necessary.

It must be clear that anyone thinking about buying a small water would be advised to look into the underlying geology. Help can be got from the Water Authority and the Institute of Geological Sciences, but you can also obtain official reports for yourself, consult what are called 'solid and drift' geological maps, and generally investigate the area. That way you'll soon appreciate how many different situations exist.

Some examples will help.

One West Riding lake had a depth of about three feet, at most. Nearby, a well was dug to take water out of a sandstone layer (sandstone often makes a good sponge). The level of the water in the lake began to drop drastically, and the anglers blamed the abstraction. What they had forgotten, however, was that in the previous winter, in seeking to deepen the lake with an excavator, they had cut through the clay bed of the lake allowing water to escape once the wet weather water table was lowered during the summer. The water in question was man-made, with a puddled clay bottom! And it had been made by man because nature did not provide water at that place in summer. The well in question did not affect the lake at all. A geological survey would have saved both problems and expense!

A similar problem arose some years ago in the Fens. After a

geological survey it was considered risky to deepen the lake because a sandstone layer passed very close to the surface just below the bed of the lake. In the event, deepening became unnecessary (the water level was being lowered by pumping in nearby gravel excavations and the water simply flowed out along the gravel bed). It was also shown, finally, that the sandstone layer held a very good head of water so that deepening would, in fact, have been of no danger as far as water itself was concerned. But, because the chemistry of the water in the sandstone layer was different to that in the gravel layer (in its trace element components), the ecology of the lake would certainly have been changed, if nothing else.

A not dissimilar instance concerned a low-level gravel terrace with a ten-foot deep, small lake on it. Deepening was required in some parts, preferably down to fifteen or twenty feet, and it was known that a dryish sandstone layer was below the lake. But how far below? And what was between it and the gravel? The last question was quickly answered – clay was found by auger. Large-scale survey maps were obtained which showed in detail all gravel, clay and rock reserves. What was of great interest was that as the sandstone bed neared the lake it was some ten feet below ground, but as it reached the lake it dipped, passed under the lake at least thirty feet down, and did not come up again for a mile or so! From the managerial point of view this was quite perfect, for deepening could certainly be done down to twenty feet in perfect safety, altering neither the water level nor the chemistry, since the lake was already partly floored in clay.

Chasing streams, whether mountain or lowland, is in some ways a different and easier matter than chasing puddles! They are rocky and irregular if highland streams, not always acid of course, but usually with less bank cover and weed growth; whereas lowland streams are steadier, overgrown and quite often with healthy weed growth, as in the lowland classic, the high pH chalk stream. Their presence is governed by only one thing, namely, whether or not the substratum is wet enough to sustain them. In chalk country, as in the limestone county of Derbyshire, there are often intermittent streams which have quite dry valleys and stream beds in summer, yet these same streams may be raging torrents in winter with a good head of trout! Much more trouble than the underlying geology are the names given to streams: becks, gills, ghylls, bournes, brooks, and so on – fortunately the map is self-explanatory!

With still waters the position is very slightly different and the name may give some indication of function or origin, e.g. dewpond. Many 'farm ponds' are simply hollows scooped out for sand, gravel, clay, or sometimes rock for wall work, and are essentially a result of farming activities. They may hold water and a good head of fish. Moats can be in the same category or may be clay puddled if the castle or house was on higher ground. In each and every case you need to know the origin, otherwise you could be in trouble. In just a few cases the origin, whether natural or man-made, could be in doubt, as in some of the Breckland meres. All that is really known is that the water may be more or less steady for some years and then all is lost quite rapidly. Ring Mere is a good example, and if you walk the region nearby you'll find any number of similar ponds either dry or wet, and all called meres. Ring Mere held big tench and rudd and the Revd. Alston caught British Record fish of both species from it. We have seen it dry and, more recently, with a foot or two of water but little in the way of fish. Obviously the chemistry is good because it grew very big fish very quickly. But you would not *buy* such a water, even if you could, and you would certainly find managing it a challenge! It is thought, in fact, that these meres originated when huge blocks of ice melted late in the last glaciation. The blocks of ice were in a sand/gravel mixture on top of chalk, so that, on melting, hollows were left behind in the dumped gravel with chalk not far below. Other explanations have been offered but all seem to agree that the chalk plays a part in running off the water, during exceptionally dry periods, through a series of sink holes. (Sink holes are solution caverns in chalk and they may be blocked with clay or periodically opened up again to water flow.) If ice played a part in the origin of these meres then they are a form of giant pingo, smaller versions of which are known in many parts of the world.

Sink holes in chalk or limestone (Derbyshire and Yorkshire) also provide small ponds, but with clearly attendant dangers! The man-made equivalents, the mine shaft or subsidence lake, sometimes provide good fishing (with the ever present worry about water loss). Old mine shafts are fished in Cumbria for tench, and subsidence lakes are quite common in mining regions whether the latter is for coal or salt. Only rarely do they form *good* fisheries, and without exception they are a manager's nightmare. It is really a question of getting a bit of fishing where one can, whilst one can.

We feel it is clear from the above just how fundamental to small water fishing and management is a knowledge of the underlying rock structure and the nature of the water supply. Having got that off our chests let's have a look at the resultant ecological factors.

CHAPTER 2

SMALL WATER ECOLOGY

It follows quite logically from the first chapter that the fundamental ecology of a water is controlled by the physical structure of the lake or stream and the underlying chemistry and water flow. Roughly speaking, and from our angling point of view, the best waters are those with varied depths: deep water over ten feet defrays winterkill; shallows provide spawning and feeding grounds. Equally important is the pH value, a measure of its acidity. Acid waters (such as some high mountain tarns and streams in peaty regions) are rarely productive, whereas a pH of 7-8 (acid is < 7, alkaline > 7) enables crustaceans, insects, and plants to thrive. There are other factors, clearly, such as the necessary nitrates and phosphates, but not in excess. The pH can be improved in many waters simply by adding lime. Similarly, phosphates and nitrates can be added, but do take advice from the Water Authority in all these matters. As we point out below, others may be involved, downstream or through-the-gravel! On one still water we have raised the pH from $7 \cdot 0$ to nearer $8 \cdot 0$ simply by putting crushed clunch (an impure form of chalk) in the margins. Many continentals introduce fertilizer in the form of farm manure or as Growmore (20 lbs per acre roughly). This results in coloured, weed-free water sometimes, but with high productivity and, often, good growth rates in the fish. But in lacking weed, apart from green algae, you can have too few links in the food chain – damage to one link can ruin the ecology completely (not unlike the situation in farming monocultures). Our choice would be to avoid manuring, if possible, but to attempt to raise the pH. Manuring running waters can be a waste of time anyway as your effort is largely lost! Raising the pH is, however, usually quite feasible.

Having assessed the physical and chemical structure you should begin to have an idea of the basic ecology of your water. Examination of the weeds and invertebrate life will reinforce your ideas and lead eventually to a full ecological appraisal. One always feels that *variety* is important: the actual percentages of each item do not matter all that

8

much, because they change constantly, but no changes will take place in proportions if most are absent!

You will notice that so far we have not mentioned the role of fish in the ecological mix. The truth is that everything else comes first: fish are the vertebrates at the end of the water's food chain. No food chain and you have no fish. At various stages in this book we'll be talking about introducing fish, and the care needed, and the reasons for it. But for the present, having set the physical and chemical scene, let's have a look at the invertebrates before going on to plant life in Chapter 3.

INTRODUCTION OF INSECTS
AND OTHER INVERTEBRATES

When nature introduces plant, fish, weed or insect life into a water to which we have access we tend to accept its arrival without question or query, eventually perhaps attempting to correct more obvious mistakes where this is possible, but on the whole remaining 'content with our lot' as they say. The reverse is true when man attempts to introduce any type of wild life. Regardless of any success, our friends or enemies up and downstream of us, or on adjoining waters, are all quite likely to band together and roundly curse our efforts. Worse, they may threaten physical or legal action. Should our efforts introduce pest or disease to a water, then you can rest assured that there will be some real 'aggro'.

So it is imperative that thoughts and actions we may have or make when we introduce insects – and in this category, as the heading suggests, we are considering water life in general – should be made slowly, deliberately, and only after certain safeguards have been taken before final release of any new life.

The reason behind the introduction of new life is either to provide food for fish, or to supply scavengers which will keep the water we are managing clear, clean and sweet, this latter reason being especially important in the case of still waters. It never has been, nor will be, necessary to introduce life into a freshwater because:

a it *might* do some good,

b it looks nice,

c there is plenty of it going elsewhere and it costs nothing,

d it has worked well on a water close by, so it should work here.

Normally it is possible to introduce pond life of most sorts anywhere, but before the physical effort is made, stop and make sure that what you are collecting does not harbour pest or disease. The only way of making sure that does not happen is to keep stock in quarantine for several days – up to two weeks if possible – during which time close observation will reveal anything that seems amiss, a situation that can be remedied only by immediate destruction.

Large ponds, etc. are not necessary for this purpose. A couple of zinc baths, plastic buckets or water butts – all are ideal, can be cleaned out after the quarantine period has been observed, then sterilised before re-use. Remember to stand these receptacles in the shade, provide weed, gravel and, in general, try to make a natural habitat so the new life will thrive and eventually enter its new abode in good health.

Here are some of the things that might be introduced, together with our observations and experiences.

CADDIS LARVAE

Food for most fish either in the larva state or later as the fly. They live on the bed or among weed, carefully camouflaged by an outer case of material drawn from the immediate surroundings. You can either drag weed from the water and pick each larvae out, or you can trap them by cutting a large bunch of gorse, sinking this in water where they are known to be. After two or three days many will have transferred to this new 'home', which is then lifted out and stripped of its catch. We have heard that old brassicae roots, especially the Brussel sprout variety, make a sure attractor used in the same way as gorse but have not tried it. Caddis travel only in damp weed or moss, over a short distance.

WATER SNAILS, LIMPETS

These feed on larvae and plant tissue both live and dead and it is this latter trait that bears watching. If you have just introduced weed into a fishery then wait until it is established before introducing any of the snail family. It is astonishing just how much they can chomp through in a very short space of time. They can be captured by dragging weed and picking them out – there are some 36 species of one sort or another in the country, so some degree of success is inevitable. The bigger species of snail such as Gt. Rams Horn, Gt. Pond Snail, with the river and lake limpets make good scavengers whilst the smaller varieties, many of which have beautiful shells, both flat and round, make fish food once established in any numbers. In the early season, snail spawn can be found under the leaves of lilies and on many leaves of the marginal rush families and this can be collected and hatched before introduction. Snails and limpets can be carried over short distances in wet weed; for a longer journey in well aerated water.

FRESHWATER MUSSELS

These eat algal food and minute animals which they syphon through the body. Small species are hermaphroditic, so large quantities need not be introduced, whilst the larger species (the Swan Mussel is the biggest – up to eight inches across the shell) are not. Sexing, we are told, is difficult! Whether they are worth introducing is a matter of personal choice. Ken has always chosen not to, his reason being that they attract fish-eating birds which will feed on them, and he is also not happy about their habit of using small fish as hosts for mussel larvae. A further point is that the smell of a dead mussel hangs around for days and days!

Collect by pulling from the bed in shallow water, where around a third of the beast will protrude, and introduce in shallow water after quarantine. We suppose that once established they could be caught and used for hook bait, but it would appear a lot of trouble for a very dubious reward.

FROGS, TOADS

These, including their spawn and tadpoles are well worth introducing on any water where they are not already established because they make a good item of fish food. Once introduced and established they become self-perpetuating, with parents returning to the water of their birth to spawn in turn. Catching tadpoles and frogspawn is something in which we have all served an apprenticeship when very young. Remember to introduce into shallow water, preferably under roots or weed so they have some cover and protection. One slight disadvantage – snakes like frogs.

NEWTS

Beautiful to watch in the water, especially the Gt. Crested newt which can grow as long as a man's hand. They feed on worms, insects, etc. and become food for larger fish on many waters. Catch them with a worm tied onto cotton (no hook) and when one takes, just lift it ashore; newts hold it a point of honour never to let go of food. Carry in damp moss or weed – never in water.

FRESHWATER CRAYFISH

We are not happy about introducing this species into a fishery. They eat insects and fish food, but themselves provide little support for fish after they grow past thumb-nail size. Another disadvantage is that they burrow into banks to make a home, which must weaken the bank. They live only in well oxygenated water and can be caught by means of

1. Freshwater lobsters. Crayfish, caught in a drop-net baited with a small piece of fresh kipper. Contrary to belief, these crustaceans do not like carrion and will only exist where there is clean, well-oxygenated water. Think carefully before introducing them – they are notorious bank-burrowers and can do substantial damage.

a baited drop-net, more especially during the hours of darkness, and should only be carried in wet moss. They do not travel well over long distances.

INSECTS – GENERAL

Too small to capture in the larvae stage, most can be attracted and encouraged to breed by the provision of flyboards. We mention them elsewhere and reinforce our instructions. Any piece of board up to a yard long will do, depending, of course, on the size of the water. Where they are to go into streams, shape one end of the board into a point to help it ride the current. Staple a long length of old line (not monofilament) to the end of the board equal to twice the depth of water in which it will be placed, and anchor out away from the banks with a brick, iron bar, etc. Renew as necessary and remember to paint the boards green on the upper surface where you want them to avoid attention from well aimed stones and branches from trees, thrown by passers-by who have little else to do.

Variety *is* important, and we do feel that it is a good idea to make a detailed list of the creatures you have in the water. Then see what is missing. Check with experts that the missing creatures would thrive in your ecological setting, and go about finding some. Do remember that the recent Wildlife and Countryside Act prohibits movement of some things. Again, check first and do not put something in the water just because it *might* do some good. The art of fishery management *is* an art, with a fair degree of empiricism involved – but not witchcraft.

CHAPTER 3

WEED

Weed is vital for most of the fisheries we are discussing. Not only does it provide vegetable food for many cyprinids, but it harbours the crustacean and insect life which forms the basis of food supply for cyprinids and other groups. Remove all the weed and you remove almost all the food. It is true that the hill trout streams (even when very healthy and not flooded with excess nitrates) may have a minimum of water moss and *Potamogeton spp* yet still hold and sustain a good head of brownies. Yet those brownies will never grow big unless they drop down river to do so, or migrate to the sea as sea trout. There simply is not the amount of food available nor, in fact, sustained warm water conditions – they are mostly *cold* waters.

When weed chokes a water seemingly to death – though this is far from being a necessary result of thick weed – it may be that some needs to come out. This is an aspect of the fishery manager's job that we'll discuss in Chapter 6. For the present we are concerned with what weed is present, assessing its value, and considering new introductions and plantings. We do not intend making this a chapter on the science of water plants as you can find this in any good book on wild flowers in Britain, but if you think back to your school biology classes you may recall a useful little concept of 'plant zoning'. Well, it matters not if you can't: in essence there is a characteristic type and shape of weed, forming a 'typical grouping' for each depth of water, from deep through to zero depth. Near the margins, yet rooted in water or mire, you have the emergent marginals, whilst from the dry land various plants encroach on the wet margins. So you have here an admixture of dry and wet-loving types, the most tolerant occurring in both. Some are even quite happy in eight feet of water as on dry land, such as amphibious bistort – even water-lilies will live on almost dry land for a summer season.

We'll start by briefly discussing the weeds *in* the water, describe the planting of them and then return to the emergent marginals towards the end of the chapter. We will deliberately start by talking about

14

water-lilies. These will grow in twelve feet of water in really crystal clear condition, but prefer water of eight feet or less. The deeper it is the more underwater cabbage they produce and a lesser proportion of floating leaves: the thinning edges of water-lily beds often indicate the vicinity of deeper water. They do not harbour that much in the way of food; have a habit of trapping silt and shallowing a water, and may be altogether too much of a good thing. They are better in clumps and patches – choose a gravel bar or submerged 'atoll' adjacent to deep water for water-lilies. Do not give them a good foothold in water under five feet deep because they are the very devil to get rid of – even grass carp don't fancy them.

Other broad-leaved pond weeds, with leaves on the surface, include *Potamogeton spp.* and, of course, amphibious bistort. These take a long time to choke a water and can be planted in small patches in water up to eight feet deep. They are attractive in themselves and do not cut out light penetration to the extent that softer weeds cannot grow beneath them: the combination of the two gives a good food larder. There is another type of what might be called a broad-leaved pond weed called the Water Soldier. This floats at flowering time, yet sinks for much of the year (Figure 2). Its greatest asset is that it harbours thousands of tiny pond snails called *Bithynia*. Strictly speaking the Water Soldier is a

Fig. 2 Water Soldier shown in its anchored phase, on its way to the surface for flowering.

'floater', as are frogbit, waterfern and duck weed. The last group to mention are the totally, or almost totally, submerged types, including Canadian pondweed, water milfoil, hornwort, starwort and water violet: these are the soft plants that anglers so often curse, yet they form the real larder in most waters.

This is not a treatise on how to identify water plants but our illustrations will certainly help. Most waters are dominated by two or three species of pond weed and soft weed, so identifying them is no real problem. However, you do need to do so. Also mark the positions of the beds on your plan of the water and try to estimate their contribution to the fishery as a whole. Do they trap silt; do they assist fish with spawning; do samples harbour a rich crustacean, mollusc, and insect stock; is the bottom beneath the weed beds 'clean' or anaerobic; does the extent of the weed bed interfere with angling too much? And so on – try to remember that weed is an asset and you really need to know whether it is doing its job.

Only when you have doubts about the last should you consider introducing new types. Have a chat with your local Fisheries Biologist. He will be able to tell you which weed prefer alkaline conditions. *We'll* tell you how to get rid of the weed you already have, should you decide it is necessary to remove them (Chapter 6). But supposing you have a virgin water, or one that has no weed growth, and you want to plant – what then?

PLANTING WEED

How not to do it is adequately described in Ken's efforts on 'Leg o' Mutton' (see Chapter 7). Willy-nilly poking of pieces of weed into the bed is rather like asking for an action replay of the loaves and fishes miracle. Water weed, either marginal or submerged, needs some careful nursery work if it is to become established and flourish – even more important, if it is not to get completely out of hand.

Much, of course, will depend on the bed of the water where planting will take place. Beds thick with silt and rotten leaves, or of fine gravel or stones, will need a different approach to those where the bed is old-fashioned, reasonably clean mud – if there is such a thing! Where there is a decent bed with little or no current passing over it then many of the underwater weeds – *Elodea*, *Potamogeton*, etc. – can be collected, washed, bunched into small bundles and after weighting with a loop of lead just above the roots, dropped into place either from the bank or, where necessary, over the side of a boat.

Moving water where there is any force of current requires more weight than a lead clip. Here the roots should be encased within a large handful of mud or pugg, either from the water itself or the bank alongside, before being commited to the deep. Experiment a little before submerging the main body of your plants; it is amazing just how much weight is needed to carry even a small bunch of weed straight downwards.

Where the bed is poor, plants must be introduced with suitable soil – loam – in which to grow. The best mix we have come across as a growing medium has been well rotted turves, left to settle for six months or so, and to which has been added a little natural manure. If you can't get natural manure then bone meal will make a passable substitute. Avoid putting composts and the various factors that go in them such as peat, spent hops, etc. into the mix. Yes, they will provide growth for your weed, but will also tend to encourage algae growth,

2. Plastic laundry baskets are extremely useful for planting pond weeds, suitably filled with earth on a thin layer of sacking and weighted with bricks near the top. Pieces of Canadian pond weed can be clamped to pieces of lead or iron.

and to a smaller extent mineral salts – which are often destructive to new plants.

The easy way to introduce weed and soil is in a container, and not by spreading loose soil on the bottom, then planting into that. Sensible containers are fibre garden pots through which roots will grow and spread or, when you can get them, small sacks and baskets which allow the same effect. Pet shops can sometimes help. Where you wish the weed to be contained, and here especially we think of the water-lily and other big-rooted efforts, then wooden or plastic boxes and tubs are needed. A neat way of planting lilies is to push the roots through a short length of earthenware field drain pipe – then drop it overboard! When the containers are to be sunk from a boat and not hand placed on the bottom, each must be weighted with stones, etc. before the load, suitably soaked, is pressed lightly into place around the well-spread weed roots. It is worth leaving a gap between loam and container rim, then filling this with gravel and small pebbles before submersion; otherwise, water will often loosen the soil to such an extent that the plants are released and float to the surface a few days after they have been set. They usually prove difficult to retrieve when this happens!

Don't just pitch containers of weed overboard from a boat. It's great fun watching them sink, but the vast majority will land on their sides or upside down. Instead, tie string – not rotproof nylon – to the box, pot, etc. and lower it gently, cutting and releasing the string under the surface once things are in place. An old, small empty aquarium pushed on the surface provides an excellent sighting tube which breaks refraction and allows one to see exactly what is going on in the deeps whilst planting.

Protection for submerged weed that has been deeply planted is rarely necessary. Plants set in the shallows do need protection, however, and our diagram of a wire-mesh cap shows the only safe protection. Marginal plants rarely need this protection and can be set out at the edge, with a brick box being set into the earth and some loam worked into it as the plant is set – a useful tip where too much root spread is not required. (See Figure 10 on page 88.)

Above all, do wash and examine every strand of weed you introduce. Loose fronds – or worse, the nightmare duck weed – are so easily introduced and so difficult to eradicate.

EMERGENT MARGINALS AND OTHER WEEDS: BANKSIDE VEGETATION

We deal with bushes and trees that overhang the water in Chapter 5. Here we consider the grasses, reeds, nettles, docks and one hundred and one other types of herbage that may line the banks and extend out into the shallows. As usual when we talk about clearance of any sort, we begin with a caution; it is all very easy to remove, but very difficult to replace anything that goes!

Wherever possible, leave as much bankside growth and weed that is placed along sloping banks; the combined root mass will ensure that collapse, and in some minor cases even erosion, can be controlled. Having said that, we admit that there is little pleasure to be seen in yard upon yard of weeds, especially nettles and thistles, along the banks of a fishery. Although they may be helpful in holding unwanted angling visitors at bay (which is often reason enough on some waters to leave alone) it is an easy task to take out the weed with one of the proprietary weed killers on the market. We have used Weedol and SBK, both within the methods prescribed by the manufacturer, keeping a special watering can for the task which we *do not* wash out or fill directly from the water in front of us. Remember that even when watering from a can there can be wind drift, and at all times remember your personal hygiene – handwashing, etc. – after use. Read the manufacturer's directions carefully.

Replanting need not be an expensive experience. Seeds of many attractive wild weeds such as vetch and poppy can be collected and broadcast, whilst a straight-forward cheap and cheerful grass seed will usually grow surprisingly well if broadcast heavily, lightly raked under the surface, and given a good watering (there is always plenty on hand). It is also possible nowadays to buy packets of seeds of wild flowers.

Marginal vegetation such as reeds, rushes, etc. can be the most difficult to control. It is a case of wade in, grasp each plant low down and pull; then, after removing it, going over the area with a garden fork to extract those pieces of root that will have snapped off. A narrow margin of rush-type weed is useful – often necessary – along a fishery. It provides protection and a spawning area to fish and insect alike, besides providing cover for the angler. It is when that narrow band spreads outwards that problems start and the only way we know of preventing that outward creep is by excavating a deep and sudden drop-off in the bed, providing just the sort of growth conditions that marginal varieties of weed do not like and cannot cope with.

3. Not good enough. Pulling merely breaks the root of most rush-type plants, leaving enough behind to grow on. This island must be thinned and a channel cut through the middle, the gravel bottom being turned over with a fork to ensure that no root matter is left behind.

1.

2.

3.

1. Common reed *(Phragmites communis)* a name we prefer to the more recent *P. australis*
2. Greater Reedmace (not bulrush) *(Typha latifolia)*
3. Willow herb *(Enilobium hirsutum)*
4. Common bulrush *(Scirpus lacustris)*
5. Soft rush *(Iuncus effusus)* possibly with the wood small–reed (to right)
6. Water plantain *(Alisma plantagoaquatica)*
7. Kingcup or Marsh Marigold *(Caltha palustris)*
8. Yellow flag *(Iris pseudacorus)*

4.

5.

6.

7.

8.

1.
2.
3.
4.
5.
6.
7.

1. Greater duckweed *(Lemna polyrhiza)*
2. Ivy leaved duckweed *(Lemna trisolca)* with rushes and possibly flowers of *Ranunculus*
3. Water starwort *(Callitriche stagnalis)* and flowers of water crowfoot *(Ranunculus aquatilis)*
4. Water starwort *(Callitriche sp)*
5. Pondweed *(Potamogeton alpinus)*
6. Pondweed *(Potamogeton sp)* and bistort
7. Canadian pondweed *(Elodea canadensis* with tendencies towards what the Dutch call *E. nuttallii)*

8.

8. One of the anglers' "streamer weeds" in this case probably a mixture of waterweed *(Elodea)* (dark colour) and water milfoil *(Myriophyllum)*

1.

2.

3.

4.

5.

6.

7.

8.

1. Watercress *(Nasturtium officinale)*
2. White water lily *(Nymphaea alba)*
3. Broad-leaved pondweed *(Potamogeton natans)*
4. Bog Bean *(Menyanthes trifoliata)*
5. Arrowhead *(Sagittaria sagittifolia)* and watermint *(Mentha aquatica)*
6. Umbellifera, probably *Oenanthe crocata*
7. Amphibious bistort *(Polygonum amphibium)*
8. Fool's cress *(Apium nodiflorum)*

1.

2.

3.

4.

5.

6.

7.

8.

1. Rich rush and grass growth around a spring
2. Mountain streams will only have rushy or reedy margins and rare clumps of watermoss locally, though tiny water meadows are easily created
3. Preparing a small dam on an overgrown stream: weed will take in the calm water behind the dam (upstream)
4. Canadian pondweed *(sensu stricto) (Elodea canadensis)*
5. Planting pondweed in a plastic basket before sinking, and weighting the root end of Canadian pondweed
6. How weed may colonise a river (in this case grass *Phalaris*)
7. A red water lily removed from the water, showing how it has overgrown and developed in its planting basket
8. Cleaning out a small farm pond: note the marginal rushes left intact

4. The beast itself with some of its roots shown. Their 'sticking power' when you are weed clearing is enormous!

Swim clearance? Most swims – and here we are talking about the bank end of the business – clear themselves. On small moving waters one tends to rove and not be bogged down in any one position for long. A deep pool may warrant an hour's concentrated effort, but nothing like the fishing time will be spent in one position as that undertaken by the stillwater man. Where swims have to be cleared it is as well to leave a band of tall herbage across its entrance where possible. It provides background cover to the person fishing in it and to some extent keeps it from the eye of the would-be poacher. Beyond that, there is no point in playing around. Clear a sensibly sized area and level the bank so that a seat can rest comfortably. If the bank edge with the water is soft, the provision of a support – such as a branch from a tree or stones and boulders – will prevent the area being trodden into a muddy morass during inclement weather.

We find that neat swims cut, cleared and kept so, remain tidy and not only give a boost to one's angling efforts but also, where more than one person uses the fishery, keep the amount of unwanted litter and general bank treading to a minimum.

The actual choice of emergent marginals is not difficult in this country because we have a wide selection of attractive types. Of the reeds, most are suitable, yet the common reed (*Phragmites communis*)

21

should be introduced with care. If your water is less than five feet deep we would not recommend it as it encroaches rapidly and traps silt. The same applies to the giant bulrush. But with reasonably deep water to hand, or the containing drop-off mentioned above, both are an asset to any water. Common reedmace (incorrectly called bulrush by many) also give good bankside cover and, like the various irises (like yellow flag) are less harsh on the tackle than common reeds. The common reed is probably the only weed that actually cuts line (and your hands if you work it without gardening gloves). No margin is complete without a certain number of other plants, and in particular we favour the water plantain and arrowhead.

Beyond the margin, but still in damp ground, there is an even greater choice available. Fortunately most of these plants — purple loosestrife, willow herb, docks etc. — arrive without help on our part. All we need to do is provide a margin that goes gradually from wet, to damp, to dry. In running water one of the nicest inhabitants of this region, the wet to damp, is the watercress bed. In stillwaters they may or may not be a success. There is just an element of chance in weed planting, notwithstanding what we said at the beginning of the chapter – sometimes you just have to try it and see.

PART II

FINDING SMALL WATERS

CHAPTER 4

OBTAINING AND SECURING A WATER

There is a sad but true story about the chap who wanted to have a small water all to himself. He was a true fishing maniac who lived and breathed the sport, and just wouldn't be satisfied until he was the proud possessor of a carp lake. It was around that time when the carp scene was all happening, and unless you had a water that produced twenty-pounders during a long, hot summer morning then you were definitely 'non-U', and unable to say 'Richard Walker' without genuflecting.

So the chap set out, scouted the four points of the compass and came up with a delightful pool tucked way in the wilds. Needed some work doing on it, mind you – a bit of drainage, the banks raising and some dredging; but before you could blink the annuities were cashed in, and contractors were hard at it with a bulldozer.

Next came the question of the fish. Yes, there was a fair head of native stock, but some new blood would certainly do no harm and again, before you could say micro-mesh, an overdraft was arranged that allowed the fish to be acquired, no expense spared, with a top fish farm tugging at the forelock and muttering pleasantries as the cheque was handed over.

Now the road journey to and from this fishing delight was a round 290 miles, not the sort of thing you would really want to attempt in a day, or rather with a day's hard fishing thrown in. So our friend used his holiday leave during the first year and stayed over two days or so at a time. The following year the fishing was better than expected, which meant that the leave was used up within the first month of the season opening, and that only left a few jolly old sick certificates on which to get a little sport.

By autumn the sick leave expired, and so had the bloke's job. His wife also departed at that time, having decided that an absent husband, out of work and in debt, was about as much use as a three-speed pike float. But the grand finale to our parable of the fish and the fisherman did not come till the following year.

Our friend took to hitching to the water, then camping out on its banks for days at a time. Somehow this upset the owner of the angling paradise and hard words were exchanged. With no lease of any sort in force, just kind words and mutual trust that had suddenly come to an end, our friend was soon out on his ear. The punch line comes later of course, when the owner let the water on a business footing at a healthy rent to a small syndicate who fondly believed that the owner had done his own improvements and they neither knew of, or cared, for our fishing friend.

Now what, you may well ask, has all that to do with finding a water of one's own? Well, we reckon the story contains just about the best advice and help we can give to anyone starting in the 'owner-driver' category of a fishery, for it illustrates all the main traps and pitfalls into which one can plunge. In fact it is true to say that in this branch of sport, as in most others we can think of, a little negative thinking at the outset can often save money and embarrassment. So at this point let's take a few hard facts and enlarge on them.

FACT ONE – CASH

If you are a couple of impecunious writers like ourselves, with just enough beer tokens at the end of the week for a chip butty apiece, then acknowledge it, and don't go looking for water in those parts of the country where you are likely to be offered a float-rocking, rod-twitching piece of water you just cannot refuse – at a fee. Chances are you will try to meet the price, and start on the slippery slope that ran through our cautionary story.

By the same token, don't seize on the first piece of water that comes your way for free – if you know that it is going to cost good money to make it into something usable. With all the best will in the world, a fishery that needs professional contractors to do drainage, bank making and other jobs – even in a small way – should be left alone unless, of course, the owner is prepared to help meet costs. Even then, be careful and read the section on Improvements and Maintenance. In short, it pays one to keep in mind that fishing waters are like mistresses; both demand time and money, and one is always loath to get rid of them when it is time to come in out of the cold!

FACT TWO – TIME

Time and land have one similarity – you cannot actually manufacture either. Find time, yes, but that is not the same as making it. So if you are going to look for a water, then choose an area which isn't a million miles away from home. Consider your travelling time as fishing time wasted and you won't go far wrong. Of course, beggars cannot be choosers and you will have to spend some time on the road.

Distance is not the only consideration. Try to remember that there is nothing more tiring than having to crawl along 'slow' roads for an entire journey, and the advantages of the main part of a journey being undertaken on better roads. It is a sad thought that nothing takes the gilt off the gingerbread quicker than the thought of a long trip before you can fish – unless, that is, the thought that one will have to make the journey repeatedly, come rain or shine, if one is going to efficiently manage a fishery.

Our thoughts run along 60 miles as an acceptable distance, time-wise at least, to travel to a water. This means a 120 mile return journey, but people differ and the distance you will consider travelling must, of necessity, be influenced by:

FACT THREE – TRANSPORT

The type of transport you possess will govern the area you can search for a fishery. A pretty obvious statement on the face of it especially if you are well breeched and own a new, or newish car. Unfortunately there are thousands not so fortunate in today's economic climate and it is worth repeating our earlier affirmation that owning and managing your own fishery is a year-round task that must be undertaken come rain, hail, sleet or snow.

So the motorcyclist is at a disadvantage in the winter months and even more so at all times when it comes to maintenance. It is possible to carry some tools strapped to one's back, but bulky items – especially baulks of timber, etc. that are often needed – are out of the question. The only solution to this situation we can think of is the enlistment of a friend with a car on a paying basis to get material, etc. to the site, or perhaps a sharing/working arrangement could be made by 'going-in' with a local angler.

Many people still have to rely on public transport as a means of travel and they are roughly in the position of the motorcyclist as far as carrying tools, materials etc. Public transport can have its uses though – for some years Ken used the Heart of Wales railway line to reach a delightful little trout stream within walking distance of a railway station. The pool, by virtue of its ecology, needed little or no maintenance other than could be done with a pair of secateurs.

A little word of warning however on the subject of public transport, more especially where buses are concerned. People soon note where you get off with your bundle of fishing rods and word will pass around the locals that someone is fishing at so-and-so. Sad though it may seem, one can bet that talk along those lines will lead to the nastiest of all modern fishing problems, the poacher, making an appearance and the modern poacher, of course, bears little or no resemblance to the true poacher of yesteryear.

So much for the negative-thinking bit, now we can get down to the nitty-gritty of searching for a fishery. Armed with the knowledge of what financial commitments may be involved and how far one is restricted by travelling, the next step is to take a very small-scale map – one of those sold by petrol companies at local garages is ideal – and on that draw a circle around your home with a radius of a single outward journey. Within that circle you can use the various methods of exploration open to anglers we describe below.

Hang on – who needs 'methods', in the plural? – we can hear you say. What on earth is wrong with just grabbing a few maps and getting on with it? Well, if you are strong at the bank all well and good, and go ahead. But a fair warning: Ordnance Survey maps work out at over £2 each nowadays and purchasing all you will need to cover the area of that circle you have drawn may introduce an ominous colour in your bank statement. No, take our advice again and start by reading the first two chapters of this book a couple more times so that you understand the basics of geology and ecology, then extend your reading.

GUIDE BOOKS

Dozens of them have been written through the years, good, bad and plain indifferent. But most are worth reading, regardless of age or even if completely unrelated to fishing and not written for the angler. Remember, you are not looking for a ready-made and undiscovered fishery to spring from the pages and hit you. Rest assured that any angler writing a guide book would keep such a find to himself.

The object of searching guide books is to establish the numbers and exact location, where possible, of various rivers, lakes, ponds etc. that may lie within your selected area. A good fishing guide book will describe active fisheries where day tickets are issued, and private waters either leased by club or syndicate. Those written for tourists and holiday makers will often describe a river or lake in relation to items of folklore or local history.

With a full list it is a simple matter to purchase the Ordnance Survey map(s) that cover the most prolific parts, and a systematic search can commence for small waters which feed into, out of, or connect with larger listed waters.

Do not despise and discard ancient guide books that come to hand. It's amazing what one can discover from them, especially those really old County Guides that nearly send you to sleep with their dryness after a page or two have been turned. Ken is indebted to just such a guide tucked away in a Sussex Library for the location of an outstanding dew pond, small enough, as they say, to spit across, that was tucked under a reach of the Downs, ignored by all and sundry. Unmarked on the Survey map, it held excellent tench – not big fish, admittedly, but up to $2\frac{1}{2}$ pounds – and permission was granted to fish for the asking. That is the ultimate in finding a fishery and it is not such a rare happening as one may imagine.

WORD OF MOUTH

As a fully paid-up cynic, Ken's reaction to this section heading was 'What word, and whose mouth?' You see, no one is going to suddenly approach you as a complete stranger and offer you a fishing water all for your personal pleasure. It's the sort of thing that could have happened in the *Boys Own Paper* and certain Enid Blyton-type stories – but this is now the 1980's. To find a fishing water by word of mouth means that you have to put out verbal feelers in the right place, in an accepted form, and keep them alive and circulating. It's just like groundbaiting a swim, more so in fact than some people realise.

Overdo the groundbaiting and the whole thing can go sour on you. If you are so enthusiastic and stir such an interest that people can talk of nothing else, especially in a country area, then locals are going to get an idea what you are looking for, or they are in possession of a rare commodity. And rare commmodities are worth good money.

Think about tactics when you start searching for a water in your area. Visit it, and try to get to know people. Use the same pub and get accepted, use the same shops in the area, and choose times to shop when the place isn't crowded out with shoppers, so that the shopkeeper may have time to chat. Stay a long weekend if necessary and get a few introductions to people who travel the area in connection with their jobs. Then, and only then, should you start your campaign.

We know that what we have said is just common sense and the sort of approach that any sensible person would make. But you would be surprised at the number of people who, if not actually cooking their own goose where a fishery is concerned at least make its water 'evaporate'. Forcefulness, brashness and general bad manners slam gates in the countryside. Loud talk about what you propose to do when you get a water and how good you would be at actually doing it only serve to create a picture of an outsider who wants to alter the stream or pond, on a piece of land that is a familiar landmark, out of all recognition – and that, understandably, arouses a countryman's 'won't' power.

Our approach has always been to introduce ourselves as mad fishermen, and as most fishermen are thought mad by the public, we generally gain an acceptance. Usually there is someone in the community who is an angler, and in a short period the opportunity of talking fishing and resources will arise. Now is the chance to tactfully find out what may be available, and who the various owners of local waters are.

After permission to fish is granted there will be time to interest the owner in a few improvements. How you progress from there is down to the skill and enterprise of the individual. Learn to hurry slowly; the fact that you are interested in your host's runner beans or his dog is more important to him than how many fish you have caught in the past, and how well you are going to do in the future.

Oh, and going back to our opening paragraph with the 'whose mouth' quote. Remember that your mouth can lose you the water you have gained with such difficulty. Shouting around its wonders and how many fish of what weight you took last Friday can either ensure poaching, the arrival of the owner's long lost nephew who is a mad keen fisherman and who has a better right to the water than you, or a suggestion that the water really is worth a 'small' fee. Be warned!

PHYSICAL EFFORT

Not such a good method of finding a fishery, at the outset at least, as one may think. It implies that you are going to tramp along and over anyone's land to reach a piece of water. This method leaves you at an immediate disadvantage, that of being a trespasser, and landowners have pretty fixed ideas about trespassing.

It is much better, in the first instance at least, to get a large-scale map of the area, make some enquiries at the local library, or from the nearest rambling club if there is one, and establish the position of local footpaths. We have a theory that the fascination of water to mankind over the centuries has led to the majority of waters being approached by a public path of one kind or another, if not right to the water's edge then at least close enough for you to make a few assessments of the water's probable worth.

But, you may ask, what is wrong with finding out who owns a water, then going round and knocking on his door to ask permission to investigate immediately? Well, nothing in the long run and of course that will be the ultimate aim. But try to avoid the direct approach in the first instance. Time and time again we have enquired of an owner whether or not we could look at a water only to be immediately told that there were no fish in it.

Now, an answer like that leaves you pretty much out on a limb. If you have not done some homework, all you can do is either say 'Thank you' and shut the gate, or push the fact there might possibly be fish – which is to start an argument that could well end up with the owner uttering words to the effect that if he doesn't know what is in the water no-one does – or is going to! Either way, you are on a hiding to nothing.

Much better to do some homework before you ask, then in the nicest possible way you can present some facts and observations. Chances are that if you play the cards right you could interest him in exactly what, if any, the fish are, and you are 'in'. It is strange, this 'there are no fish in it' syndrome. There are thousands – probably more – small waters in this country which people fondly imagine are devoid of fish life, but which are often bursting with fair stocks, and often some good specimens. Not only that, but the waters that may be devoid of stocks but which with a little management could be made productive must be even more prolific.

Walking is the penultimate of fishery finding, to be undertaken when one is confident that the slice of water is worth considering and viewing. But there are areas in this country where walking would take you a month of Sundays to cover or even reach a worthwhile water, and where transport is necessary over some really rough ground. Ken immediately thinks of Wales, where there are thousands of game fishing streams unfished and often unexplored by an angler. To reach them you have to leave the road and take to the tracks, often along dusty miles of broken stones and ruts before you can even reach a farmhouse, let alone a bankside.

This is where the motorcycle comes into its own, and investigation can quickly and cheaply be made with its help. The beauty of those little mountain streams is that they often need little or no maintenance, so the motorcycle can be used exclusively as a means of transport without the worry of carting tools and materials. But note that we said 'before you reach a farmhouse, let alone a bankside'.

Scrambling, grass track race practice and general high jinks that leave a field scored and rutted by tyre marks will bring an obvious answer to a request of any sort. The final approach must be on foot.

WHAT'S WORTH CONSIDERING?

Answer, anything – but considering is not negotiating for, or working on a fishery, at least not until an awful lot of pros and cons have been investigated and balanced against each other. And the pros and cons of this section are not centred around the ecology of the water, at least, not the small-print things; at this stage it is the big items that tend to be overlooked in the first excitement of discovering Valhalla.

Let's imagine we have found a delightful length of stream some distance above a main river, holding fish, and to all intents and purposes 'ripe for development' as the advertisements say. It has fulfilled the obligations of doom and gloom we spread earlier in this chapter, and appears, on the surface, to be a sound proposition. What could there be that would mar this delightful dream?

First there is the question of access. Now, if you drive or walk off the road and reach your stream remaining the whole time on the same owner's land, all well and good. But things aren't always as simple, and often you need to cross the field of another to reach where you want to fish. Or there is a long way round and a short cut to the water, again through someone else's land. Perhaps there may be a footpath to the stream but you will, during the course of work, need a track along which to bring materials, and the track lies on an adjoining owner's land.

All very theoretical you may think, but these are the sort of things that can and do happen and which really cause problems and spoil your peaceful fishing. So take a very good look at the map, look at the water, look at access routes and check ownership. If you need to cross someone else's land at any time you will be better to sort the matter out before you commit yourself in any way. Not all landowners are good neighbours with each other. In farming communities, the opposite is more common!

How much background history do you know, or can you discover about the water? We are concerned with things like flooding in the winter, and drying out and water abstraction during summer months. They all happen; floods carry away banks and erode work that has been undertaken, which must be expected occasionally, but for all you know it may happen every year which is, in short, a different kettle of fish.

Drought conditions affect all waters, but if the one you are considering dries out, albeit only one year in six or so (a peculiarity attaching to some waters) then you are going to waste time and money in improving fish and weed stock. Worse is abstraction, which can play havoc each year with small streams. Ken had such a fishery in

Hertfordshire, along the Lea Valley where the Water Authority allowed Market Gardeners to abstract water for field irrigation during the summer months. Grand winter fishing – but a waste of time taking a rod to the water in summer and eventually he gave it up.

The final consideration, the one that causes most heart searching and not a little hardship when one is not scrupulously honest with oneself at the outset, is that of actual physical labouring which may be required either from yourself or from paid professionals either to improve, or get a fishery off the ground.

Unfortunately this is not one of the things you can commit to paper, say, as a table or guide. One cannot say that X amount of work is needed and Y amount of money or effort will be required.

To date we have considered the angler as an individual, which most are, who is looking for a fishery of his own. For this reason we have slanted our advice along a low-profile cash base and feel that the vast majority have to accept that situation. Now, it is possible to carry out quite a lot of physical work. But if it is done on a one man basis then it is going to take one hell of a lot of time – time you will not be able to spend fishing. Often, too, the elements can overtake one and the incompleted work of summer can be swept away by winter floods.

Somewhere there is a break-even point in running one's fishery, no matter how small that fishery may be. Only you can decide whether the work can be avoided or ignored, and whether you are physically up to it. We have both attempted waters and given them up for one or more of the foregoing reasons in the past. Had we been honest and realistic over work which just had to be done in the long run, we would feel better about it all today.

But two or three anglers banded together to run a fishery can make a world of difference when it comes to maintenance and construction work. We shall have a little to say about small syndicates, or rather some of their problems, in a later section, but a very few select friends running a water big enough to accommodate them can tackle mountains, and even find cash for some professional work. We also find that in such a situation there is always a member who knows someone who knows someone else....

Be realistic and utterly detached when you are considering a water. It sounds rather wet, but the best advice we can give still centres around the pros and cons. Set them all out in writing, cost them financially or physically, and weigh them against each other. An honest picture can only emerge, either on a single basis, or as a possible syndicate. How many members make up a syndicate, for example?

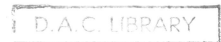

PERMISSION – AND REFUSAL

We think that we have made our attitude towards obtaining a fishery of one's own pretty clear. We hurry slowly, do our research on the water, then ask for permission to fish. Only after we have fished a water for some time in fact do we consider our next move, that of establishing the fishery as our 'own'.

How do you set about crossing the barrier between casual permission and management? You can start making slow improvements, trading on friendship and using excuses, if you are challenged, that what you have done is an improvement and of benefit to the owner. But such lip-service wears thin in time. We all, eventually, become aware when the tail begins to wag the dog, usually through a third party high-lighting the cause.

Life is full of compromises and unless you purchase a fishery you will be unable to rip things apart and start again – and not even then if the Water Authority think otherwise. So improvements must be those which meet with the owner's approval and if you can arouse the owner's interest sufficiently to take part in the work then you are on a winner.

Our approach is to work out a series of jobs that will benefit both parties. Sensible jobs, things that the owner can see will need doing now, or in the near future. We also make sure we can illustrate what improvements these jobs will bring both aesthetically and piscatorially. From there it is a matter of waiting the right moment to put the case forward.

Usually the plan works. Occasionally it doesn't and refusals are difficult to counter. We have talked to a lot of our friends on just this point before putting pen to paper, and analysed their, as well as our own, past reasons for refusal from owners. Quite frankly, most of them have been triggered by anglers jumping the gun, wanting to do improvements long before a position of mutual trust has established itself between angler and owner. Quite a few have arisen from thoughtlessness. The angler has wanted to grub out trees and bushes, or drastically prune them and has expressed himself badly on the point. Some have arisen, we are sorry to say, from selfishness on the part of the angler who has deemed it fit to tell the owner that he is doing him a big favour by improving his poor water. Think well on these points before you make your approach.

Syndicates will, in most cases, be tied to a lease and what can and cannot be done will be clearly stipulated on that document. Where a syndicate runs on goodwill without fee or lease, remember that the owner does not want to be pestered by each member with ideas and improvements every time they visit the water. Elect one member of the syndicate as spokesman and let him do the talking. Not only will matters be quickly resolved, but misunderstandings avoided.

LEASING

This is the Big Boys' League, and if you are contemplating a lease on a water then get yourself a solicitor as soon as possible and brief him. Note well that we said brief him, for he is there to do as you want, and whatever you feel should be put into the lease or removed from it is up to him to negotiate. Not least of his job – the most important probably – will be to ensure that whoever offers the lease is entitled to do so. Ownership of land does not necessarily incorporate the ownership of fishing or sporting rights, and those rights may already be assigned elsewhere.

Try to get a long lease for obvious reasons. Leases that must be renewed each year create legal expenses and do nothing to encourage long-term investment or work. When a lease is for a period exceeding three years it should be by deed, and that is the sort of arrangement you, or your syndicate, want.

Many other things need sorting out before the lease can be signed and your solicitor will protect your rights. He may not be a fisherman, however, so these few suggestions may serve to remind you on what may need attention. Obviously parking facilities and rights of way, both of which can become volatile issues especially in the long term. Allowances for improvements you make should, if possible, be compensated for and you want some clear idea of exactly who will have the right to fish. Often the owner may want to use the water himself, or stipulate that his Uncle George can fish on his holiday. If you cannot get that erased from the lease, it is a good point for haggling over on the amount of rent to be paid, and that, also, must be stipulated.

Those allowed to visit the water should be stated and named if possible, which allows you to concentrate on a clause stating action on poachers that may be found. If you cannnot think of other problems, may we suggest you look for ideas in the 'A - Z' of problems at the end of this book? As an example, the entry on Abstraction.....

A word of warning. Under certain circumstances angling waters are liable for rates valuation, and rates may be payable to the local Council. Check whether this is so, and if it is, find out what the rateable value is and whether the lessee or the lessor has to pay it. More than one person has had a nasty shock.

To give you an actual example of what can happen, we'll recall Barrie's experience of several years with a carp lake syndicate. The water, of some four to five acres, became available after the 1963 freeze-up killed all the fish and no local angling club would take it on again. With the help of a solicitor two of us negotiated with the farmer a ten year lease at a fixed rent p.a. (It was, in fact, £50 if we remember correctly.) Frankly, a lease under ten years is not worth considering for a water that needs a lot of work and/or stocking. And this one did. We also had right of access, exclusive fishing rights (owner excluded), total control of all bankside work within ten feet of the average waterside, and the right to remove fish at will by rod and line angling. The syndicate involved ten anglers and the fishing flourished and was very pleasant.

Eventually the farmer sold out and the new owners moved in, but did not seem to realise that they had no right to fish, let alone know that a syndicate had firm rights. When the owner was found fishing we did not stop him but carefully pointed out that he would be doing so only as our guest. At a later stage he began to do bankside landscaping and within two days was stopped by court order. Make no mistake about it, a good lease is a powerful weapon. However, having great sympathy with his position – after all, he lived there – we negotiated a new lease giving the (now) nine members of the syndicate *free fishing for life*, with total access at all times, the right to remove fish by rod and line, but giving the owner the right to shape the banks and the growth thereon. A secondary advantage to the owner, in this arrangement, is that as time goes by the syndicate members will leave, stop fishing, or die, so that in the long term he and his family will be back in sole charge and possession. Both parties considered this a fair arrangement. A clear risk to the syndicate members, of course, would be that the owner could make life so unpleasant that everyone would leave! Well, to cut a longish story short, the gentleman who owned the water did not take this line at all.

The clause 'the right to remove fish caught by rod and line' not only enables you to eat a fish, but, with Water Authority permit, to stock a new water should you buy one. That's exactly what has happened.

BUYING

This is the ultimate, and one we have just achieved as members of a small, progressive, and, thanks to past treasurers, well-heeled village angling club. We paid £19,000 for fourteen acres of land, half of which has an old, established gravel pit with good depths and a good head of fish. The costs of buying a water, be it still or moving, have soared this last decade: a friend of ours had a chance to buy an eight acre clay pit some twenty years ago, for £300, and turned it down as too expensive.

The prices of today are today's market value: they are real and realistic. If the price is too high they do not sell. Ours, bought in late 1982, has already increased considerably in value as we pen this only a few months later. So do not be put off initially by seemingly high prices.

A solicitor is a must, as with leases, but here his task is slightly different and in some respects easier. For example, rights of access are usually included and clearly laid out – if not, no-one would buy. But, like the owner in the previous example, do make sure you have the fishing rights! (and, whilst you are about it, the shooting rights, mineral rights and so on). We have all these, together with a good mineral resource, and seven acres of land upon which to play. It is quite astonishing that once members realise they *own* a water, that age old problem of all clubs, the so-called work party, completely disappears. But everyone on our water works hard and often: the problem is to hold them in check.

There are, of course, many ways of financing such a buy, but basically the club has to have a reasonable bank balance to help convince the manager that a loan from him would be repaid. There is no point in a club using all its capital: all you need use is enough to convince the bank manager you are serious, because he'll not loan you 100%! Of course, he'll quickly see that a good water will yield a sizeable income from day and season tickets and from matches. If you are prepared to issue day tickets (and our club has always taken an enlightened, almost too philanthropic attitude to other anglers) there is a possibility of grant aid, for example, from the Sports Council. A grant is a grant and needs only to be repaid if you sell the water or renegotiate on the agreement to make day tickets freely available (depending upon what the water, as a fishery, will stand – this goes almost without saying).

So there you have three sources of money: your own; a bank loan, and a grant. The next step is to get rid of the bank loan as soon as you can unless you are a very rich club, in which case it's probably cheaper to keep it as long as possible. One way to get rid of it is a very obvious one. You need a hardcore of hardworking men in a club of this kind, including experts (builders, carpenters) as well as fanatical anglers who, although they normally never do a day's work in their lives, will work unstoppably upon a project of their own. Therefore, get such characters to sign up as life members at a sensible fee (at least £200 obviously). They get, in effect, a water of their own, for life; can (and do) get really involved with it; the day and season ticket holders also benefit much more quickly than they would have; and the club treasurer can pay back the bank quickly. In our club we had aimed to pay off the loan in about one year, having been given eight years by the bank, but in fact we cleared the bulk of it, by this means, in five months.

You can, of course, use all the usual money-making schemes such as dances and discos. Be warned of one thing: do not make your plans too public at an early stage, and the Committee should keep strictly to itself, so far as its constitution permits, its financial planning (or at least the detail thereof). In our club we had a few low key plans for the water, and were very open about everything. Some locals got wind of them and we suffered some very under-the-belt opposition even to our purchase of the lake, let alone our plans for it. We won the day but not before learning who were our friends.

The enormously satisfying feeling of owning your own water is something you can savour for months. There is nothing the objectors can do once you have it and if you have planned well and got your facts right, you call the tune from then on. You join that section of your club to the A.C.A., if it's not fully covered anyway, and you contact the A.C.A. to let them know your position and any danger points with respect to possible pollution. They'll come and inspect if you ask them. Now you can settle back and think out your long-term plans, which we cover elsewhere in this book.

CHAPTER 5

MANAGEMENT AND IMPROVEMENT

THINKING

Work on maintaining and improving a fishery never, never ceases. There will be times when you may think you are on top of the job and that you can sit back and relax for a few seasons, then along will come an unexpected flood to scour and collapse a bank, or a drought which encourages water weed and bank growth to climb to chest-high proportions.

Added to that must be the knowledge that no two waters are ever alike and rarely, in our experience at least, conform to the text-book descriptions of management and treatments. In fact, many things in management become a compromise, and often of experiment, the results of which may appear doubtful at times.

An example is the dam which one builds to try and raise the water level. This little job automatically reduces the speed of the current, which allows water to drop its silt into the now near-still water above the fall. This will, in time, create a shallow that can warm up and eventually become a weed bed. Given more time this will choke the stream, and ...

Yes, of course, we have taken things to an extreme. But it happens, and the only way to try and avoid some of these 'knock-on' effects is to do nothing. Yes, that's right, nothing of any sort in the way of management on a new water for twelve slow, long, agonising months. During that time, haunt libraries and bookshops and read every publication you can buy or borrow on management, then re-read them several times more.

Next make a book of your own, a diary, completed on every occasion that you visit the water, whether you fish it or not. Such visits, if you are going to achieve any success in management, should be on a twice-weekly basis at least. Note every single thing you can in your diary. Each insect, bird, fish, mammal, weed, bankside weed and clump of herbage – everything that is both usual and unusual, good, bad and indifferent, and what happens to it all in the end.

Add to that diary the weather at each visit and a review, if possible, of what it has been for the preceding few days. Someone locally will give you a good idea of that – there is nothing like asking the Englishman what he thinks of the weather! Use a water thermometer and record temperature and try, at two or three selected places, to regularly record the depth of the water.

There is a great deal of hard work contained in that little project, but it will be of little use unless you can marry the information into a map of the water. Not an Ordnance Survey map, but one of your own making and not necessarily to scale, but one which shows every nook and cranny along, and for some distance back from, the banks. Trees, bushes and weed beds are obvious items for inclusion. But of equal importance are drains – even small field drains – and ditches, large underwater obstructions, and the composition of the banks at various places, especially areas of marsh and stone outcrops.

Don't be frightened to give various areas or stretches of water uncomplicated names. You will be surprised how difficult it can be to identify a certain feature even on a small water when you are writing up the diary or, if the water is on a syndicate basis, to discuss work that is to be done at a certain point. The names can be as daft as you like providing they graphically describe the place. We plead guilty to Chub Straight, Hazel Arches, Clay Bottom and Moses Place, to name but a few.

Photographs taken during that first year are invaluable. No, not glorious works of art but sensible black and white or colour shots of each area taken through the months, which will show seasonal changes. In the long term they will also prove to you exactly what has happened after each change you have made, or each job you have done badly. Use the same vantage point when you repeat each photo, and make sure that you have some sort of index system that shows date and time as well as the exact point on which it was recorded.

Following on those long twelve months of waiting you will, with map and diary plus photographs before you, be able to analyse the behaviour of the water, assess its wild life (both insect and fish varieties, plus bird and mammal) and then decide on the management jobs that need to be done. We have a maxim which we apply at this point of management, one we took a long time to learn, but which we pass on to you as being necessary when work of any sort is contemplated.

In its simplest form a water, either still or running, needs oxygen.

Oxygen provides food and life, and without it the water dies. The only way it can be introduced and held in the water is by atmospheric action, i.e., wind, rain, etc., or by photosynthesis; a method whereby the sun causes plants and weeds to shed oxygen into the water around them. With a few odd reservations, we therefore recommend that unless the work you are going to undertake will improve or assist that action, either in the long or short term, it will be best left alone.

You see, we anglers have many faults and vices, not the least of which is that we like to see a water as being 'tidy'. Neat, trimmed, snag free, uncomplicated banks to help in casting and to land a fish are high on our list of priorities and we tend to rush in and carry out these 'improvements' straightaway. Now, fishery management just isn't like that and cosmetic surgery along the banks, or landscaping jobs made simply for their aesthetic appeal are but two of the easiest ways we know of ruining a water in the long term. Each and every job that one imagines could or should be done MUST be examined against our Golden Rule of what effect it will have, or could produce, on the oxygen content of the water.

There are two further things one should consider before rushing into work on the fishery. First of all, there is the Water Authority, who have a vested interest – sometimes reinforced by the law – in what you are doing, or intend to do.

We all have our thoughts on Water Authorities and their powers. We have, on umpteen occasions, voiced our thoughts to various Authorities on odd things, and probably we will do so again. Equally, we have also had a great deal of help and advice on the theory and practice of various problems on small waters that we have taken on. Some of those experiences are set out in our chapter on Case Histories, and you must, if you are without experience of your own, derive a few thoughts from those.

As in everything in life, we find that a bull-at-a-gate approach to Water Authorities will produce a large and daunting NO. An approach asking for help, outlining what you want to do in a graphic, well balanced and carefully illustrated way usually results in help, advice, and if the worst comes to the worst, a means of compromise.

You will do well to remember two jobs which definitely need permission from your Water Authority before they can be tackled. One is in constructing weirs or dams, the other is in introducing fish into a fishery.

The further consideration one should make before undertaking maintenance or making improvements is the one we find will lead to the most awkward and difficult problem in modern fishing; that of the poacher and vandal. There are times when it seems that they are watching your every move, waiting to move in as you move off the water. Of course, the miserable thought is that the harder you work on a water the more your efforts will show and the more people will realise that the water is being improved; therefore, the chances of unwanted visitors increases.

If the owner lives on the water, or you are working with a syndicate where one member is a local, or the water is buried so deep in the countryside that it won't attract attention – all well and good. If your water is close to human habitation and visited either by yourself or others on a weekly basis only, then take a careful look at every job you undertake on the fishery and try to make the adjustments or improvements as surreptitiously as possible. More important than that, don't perform jobs that invite vandalism, the other 'sport' that seems to be on the increase. Concentrate on performing each job so that it has an in-built second-hand appearance, and so escapes notice. It is not quite as difficult as it may seem!

TOOLS FOR THE JOB

Big or small, the jobs you are going to tackle will all need tools and a few words about some of them won't go amiss here. We have made a point of not purchasing all our implements from new. They are going to be used in wet places, often hidden-up under a hedge or in a ditch to avoid unnecessary carting around between visits, and generally remain out in the weather. But second-hand or not, we look after our tools and clean and oil them at regular intervals. Here is a list of them together with a few comments, and there is a photo of Ken's collection as well.

SPADES AND FORKS

The garden varieties will do, and we find second-hand to be as good as new. The spade will be good for most bank work, whilst the fork will be necessary to lift and split-out the root systems of some species of water weed. Two forks on those occasions are better than one, both being inserted back-to-back into root clumps then levered against each other. Occasionally a fencing spade – one of those things where the edges come towards a point – has a use, but not enough to consider its purchase. There is a variety of fork which has its tines turned at right-angles to the beginning of the blade area and these are invaluable in hooking out emergent weed or water-lily root complexes.

5. Tools for maintenance work. Some of those that Ken uses. Note the galvanised buckets; plastic ones won't stand the strain and soon fall apart at the handles.

SAWS

We use one variety only, a 30 inch Bow-type (see photo) which is light to carry, easy to use, even under water, and possesses the advantage that the blade can be renewed in minutes. Finding a saw-sharpener, let alone paying one, is a nightmare job today, so take care to protect the saw edge, when it is not in use, with a piece of plastic sacking. We consider this item is a must and should be purchased new.

HAND SICKLE

Necessary to keep the banks trimmed, this tool, which rejoices in a variety of names in different parts of the country, should also be a new purchase. Choose one which is light in weight and take care to keep it sharp and the cutting edge covered when not in use. When in use, carry a slip stone with you to maintain a keen edge.

BILLHOOK

There are a few uses for this tool in cutting sapling-sized timber and sharpening points to fencing posts, but it is possible to manage without it. Worth waiting for a second-hand model. Like the sickle, it is useful for cutting and simultaneously stacking brambles and brushwood. Its longer handle makes it, perhaps, more useful in this respect than the sickle on occasions. As with the sickle, carry a slip stone to renew the edge.

AXE

Of more use than the billhook. It will split posts for fencing, help to root out wood stumps and the back edge of it can be used, in an emergency, as a sledge hammer. A new one lasts a lifetime and has uses for supplementing the household fuel bill. Never leave a rough-hewn finish with an axe. Skilled use is always indicated by a cleaned up end to a stump of tree, for example.

BAR

A heavy tool with somewhat limited uses. It certainly makes post driving an easier job, and is best hired, when necessary, from a D.I.Y. shop together with its counterpart, the ...

SLEDGEHAMMER

There will be occasions when the gentle persuader will be needed. Try to assemble several jobs that will require its use so that you get the most from a day or two's hire.

CHAIN SAW

This can be hired. Experience cannot, so make sure you have a crash-course on how to use it, together with an outline of what it will *not* do. We feel that using a chain saw is a job that requires two people, if only to cope with a possible emergency. When a chain saw is hired only the hirer should use it. When a chain saw is owned by one of the members only he should use it. Never lift a chain saw to shoulder height.

BUCKETS

Two are needed for shifting silt – buy the metal variety, not the plastic ones which break away at the handle when used for scooping.

SECATEURS

The maid-of-all-work; the tool it pays to carry at all times when you are at the water. It slips into a pocket and always seems to find a task or

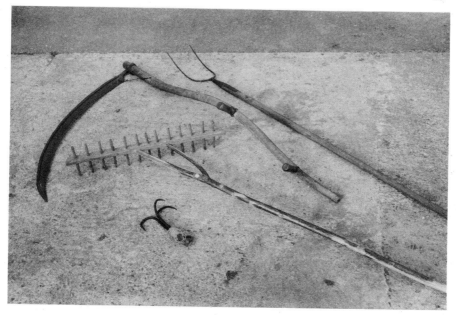

6. Weed cutting implements. Nothing very exciting, but sufficient to handle most weed problems on a small water. Compare the professional approach in Photo. 34.

two along the bankside. Worth investing a few pounds for a heavy duty pair of good quality.

SLIP
Last, but by no means the least. A small slip stone will keep the cutting edge on all your tools (apart from the secateurs) and reduce work to a minimum. Take care with it for they are fragile and smash when thrown to the ground.

So much for the tools and now for a few general words about:

TIMBER AND MATERIALS
Some timber will probably be cut-out from the work that goes on along the banks, but not all timber is suitable or usable, especially when one

is looking for stakes, etc. It will occasionally be necessary to buy rough timber, and the best source for this is at a demolition site. Often, too, a farmer may be able to raise a bit of suitable wood from his barn.

We would like to voice all the usual caveats about well-seasoned timber, but such a commodity does not seem to be available today. Select the best you can and clear-Cuprinol or Creosote it, a task which, apart from protection, will tone-down and keep the natural colour of the wood, so helping to divert attention and keep down vandalism.

Nails in various sizes, plastic sacks from a local farmer, wire and wire netting will be required in odd quantities, together with a good roll of strong green nylon garden twine for which one can find a thousand good uses.

WORKING

So to the work of maintenance and improvement. In this section we list the various jobs one can do, together with reasons why they may be needed and some of the effects they can bring. In Chapters 8 to 11, we set out various types of water and list their inherent faults, then suggest remedies from this chapter that might be used to correct them. By this means we hope we provide a cross reference of what to do and the easiest way of doing it.

DAMS

Something of the small boy remains within us all and most of us itch to make a dam of some sort or other across a water. The reasons for making this type of construction should be one or more of the following:

1. To provide holding pools for fish and insects, both above and below the fall, and to give deeper water in holding pools that are already in existence and well established.

2. To prevent earth eroding from 'fast' stretches of water, and stop the erosion of earth from around the roots of water weed in those regions.

3. To provide oxygen – this sometimes on a casual basis during those long, hot summer months that we very occasionally experience.

4. To remove accumulated silt downsteam of the fall.

5. As an occasional weed stop when weed cutting is in progress upstream.

6. To control the level of water held within a still water, i.e. pond, lake.

7. (Left) Wire netting fence. This one is made from pig wire; stronger than the usual chicken mesh but equally useful year-round. Heavy debris during winter months rarely damages this type of dam.

8. (Right) Simple log dam. This one is only six inches high, but it is sufficient to encourage insect life and provides a useful pool along a trout stream that holds fish. This is its third year of operation, the log in question being a stout oak.

If you are looking for, or thinking in terms of a large-scale job holding back six feet or more of water and offering boards or paddles with which to control the level and flow of water along the fishery, start first, as we suggested earlier, with permission from Water Authorities and landowners around you. There are also some legal ramifications that centre around Land Drainage Laws and it would be as well to look for a solicitor to help you over those points.

Once those hurdles are cleared, the work you want is still hardly likely to be within an amateur's grasp. Quite frankly, we have never had to envisage a task like it in the many waters we have managed or maintained, and cannot recall when we have wanted to lift the level of water by means of a dam by more than nine inches – and then only on a temporary basis. Large dams today are usually employed to hold back a stream and flood an area that will form a small trout lake, or form an escape where such an artificial water is excavated. In that league, professional work is the cheapest solution.

Small dams – barriers would possibly be a better term – that may be needed can be made from stone, where that material forms most of the stream bed; suitable alternatives are wood, or wire netting.

STONE DAMS

Most of us have played at them at some time or other. To be efficient, however, they should be three or four times as broad across the base as they are in height. To withstand water pressure they should curve very slightly backwards i.e. upstream (see Figure 3) and should have a dip in the centre allowing an easy overspill, away from the vulnerable edges.

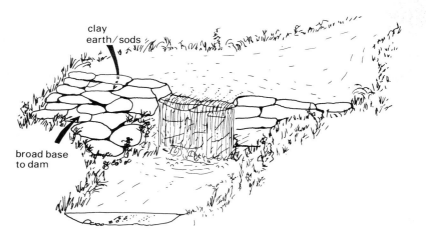

Fig. 3 Stone dam, with broad base, and earth, sod, or clay on the upstream side. The pool acts also as a silt trap and must be dredged or flushed at intervals.

Those edges, the points where the dam fits into the bank, are always the weakest spot in construction and water invariably tries to creep behind them, washing out the earth and demolishing the whole structure. Our remedy for this is to cut into the bank, insert a plastic bag – one of the heavy fertiliser jobs from a farm – and seal this in with earth, leaving the loose end out and into the stone wall of the dam.

Depending on the width of water to be held back these dams take little time to construct and maintain. They carry away easily in a flood, and by so doing won't cause a hazard. If there is one disadvantage to be found with them we would say that when placed across a water in or close to a built-up area the public at large cannot resist either enlarging and improving on them, or destroying them altogether.

9. Stone dam on a mountain stream. This is rather higher than one would usually build, and was erected during the course of a very dry summer. Subsequently, a storm levelled it – the very reason for its existence, of course.

49

WOODEN DAMS

The natural material to use if stone is unavailable. It does not, however, make such an efficient construction as the other types we describe. Trouble comes when it gets washed away in a flood and two or three whacking great pieces of timber go charging downstream. Naturally this will eventually jam somewhere and after a short time divert flow, damage banks and generally cause mayhem – perhaps, on someone else's stretch of water!

Ken has worked out a system for a wooden dam that only uses one piece of timber; a substantial branch or perhaps an old wooden rafter long enough to reach from bank to bank. This is set into the sides at the height to which water is required to be raised; the space below the timber is then filled as far as possible with mud, gravel, etc. and then this is covered with plastic fertilizer bags, or a length of heavy polythene sheeting that will reach across the stream. The plastic is laid so the edge rests over the lip of the wood.

Now more earth and gravel is heaped onto this 'wall' to hold the plastic in place and within a short time the dam will hold, and even improve. The great advantage of this method of construction comes when a sudden flood takes place. The sacks and debris that form the dam's wall will soon wash out, leaving a single piece of wood that will allow water to pass both under and over it, especially where the wood has been secured by a wire and peg anchor to the bank at one end.

Fig. 4   Simple log dam essentially of use only on the smaller streams: shown cut away at right to illustrate structure.

10. (Left) Damming a feeder stream with the log and plastic bag dam. Note how small this little feeder is, compared with the next photograph.

11. (Right) Taken half-an-hour after completion, this photograph shows how already the stream is swollen some distance back. Within a few days there will be a significant increase in the head of insect life, and after a fortnight (providing flooding does not occur before) the dam can be lifted and the insects released down to the main stream, some thirty yards below.

Rarely have we found the single timber strut to wash completely away and sacks that do so can soon be retrieved without causing damage.

A very simple temporary dam (Figure 4) which we first saw used effectively on the upper reaches of the Yorkshire Derwent consisted of four posts hammered into the bed. Dropped between them were the required number of poles which could be added to, or subtracted from at will. Naturally, the ends had to be secured into the banks and also into the river bed, otherwise water would leach past. Such dams are best used on waters no wider, at the dam, than ten feet or so, and would definitely come under the heading of those for which permission would be needed for construction from the Water Authority – more especially where migratory fish run the stream.

WIRE NETTING DAMS

Our favourite. They are cheap, simple to erect and to collapse, attract little attention and can be laid across quite wide stretches of water. Stakes are driven into and across the bed where the dam is required, and then small-mesh netting is laid over the stream against the supports, with the required height raised and secured to the stakes by staples. The remainder of the netting, and there should be several feet of it, is laid flat on the stream bed and covered with stones, earth, etc. until it is well and truly weighted down and unlikely to move.

Within a short time debris carried downstream will lodge against the mesh and this will block, causing the level of water behind it to rise. We usually hasten the process by soaking and wedging straw or grass at either edge of the netting at its junction with the bank – an area of slack water which takes longest to fill with weed, leaves, etc.

Strangely, this type of dam seems to meet with little in the way of vandalism, though we do, where possible, erect them at the natural junction with a hedge or fence, so that the whole thing seems a natural extension of a boundary.

Those are the three dams most easily erected and which we find are generally accepted, on a casual basis, by the Water Authority. Again we have a few golden rules to follow when we build any or either of them, which read as follows:

1. Keep them as short as possible. Long dams are the devil to make and the easiest to break.
2. Choose an area where the banks on either side are high, firm, and in good repair.

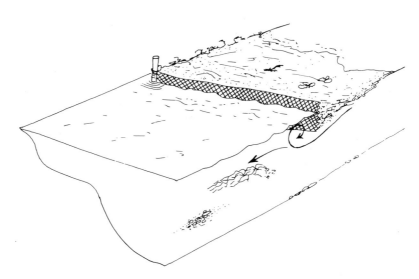

Fig. 5   Weed trap of netting or wire, used in surface layers and crossing the full width of the water. A raft builds up behind, which gives shelter to fish, but which can be cleared at intervals using a fork with its tines bent at right-angles.

12. (Left) A simple silt trap. This one, at Leg o' Mutton, was constructed from a stout log and two sheets of corrugated asbestos, reinforced with the odd scaffold pole. Upstream to be excavated, and stones below the fall ready to absorb the eroding affect which the overflow will bring.

13. (Right) One year later, the silt, trapped above, can clearly be seen and is ready for excavation. Some of the chalk blocks, which were laid in batches, are in the foreground.

3. Make sure the dam itself dovetails tightly into the bank. If it does not, water will filter around the ends, eroding the banks away and defeating the object of the exercise.

4. Make sure that the dam face is at a dead 90° angle to the water's flow. If it is not, you will divert the water outwards from its centre run, usually into one or both banks – with dire consequences!

WEED TRAPS

Not dams in the accepted term. Weed traps are placed on the surface of a water and they hold back debris of every sort that descends from upstream. They are especially useful during the spring and summer months, when they prevent the wholesale spread of some pest growths, such as duckweed.

The stop is a fairly hefty baulk of wood that is slightly longer than the width of water to be covered. It is floated in place at a slight angle across the stream (see Figure 5) and secured by a rope, chain etc. at

53

either end. Enough slack is left in the 'anchor' to allow the stop to rise and fall with the water conditions. Several logs can also be attached end to end and six-inch width netting can also be used, though the latter lacks strength under pressure. Naturally, it will all need some attention. Accumulated weed and rubbish must be frequently cleared and carted well away from the edge of the bank before it is destroyed.

BANK WORK

Problems with the banks themselves, not the bed of the water, can be some of the most demanding and worrying. Action of water against the banks will bring constant wear and tear and this, combined with extreme weather conditions – the winter of 1981/2 is an example – can bring sudden destruction on work already completed. The only chance one has in this part of management is to watch and anticipate, working on the principle that protection is better than cure.

EROSION AND DEPOSITION

This dynamic duo perform their mean and ugly work literally at every turn. Current will hit one side of a bend in the stream, eroding the side of the bank, usually below the surface. Eventually this undercutting will cause a whole slab of the bank to fall into the water, altering the course of flow and providing silt problems further downstream. Obviously the frequency and amount of damage will vary with the force of the water, and this provides a clue to the treatment; you must reduce the strength of the current which hits the bank, shielding it from damage. (See Figures 17 and 18 on page 115).

The cheapest shield will be timber and brushwood secured against the threatened bank, both above and below the point of impact. The padding must be very firmly secured (to the banks) by ropes and stakes, otherwise it will vibrate and rub against the earth behind and, in effect, cause its own erosion. A spin-off from this type of treatment is that the rough wood suspended in the water will act as a natural holding place for fish, and a sure place for you to lose a 'good 'un' at least once a year.

Where the water is so small that a branch cushion would cause a blockage it will be necessary to resort to wire netting, held against the bank with stakes driven deeply into the bed. Any space between the netting and the bank should then be packed with timber, stones and brushwood. This neat job is best undertaken during the summer months when the water is low, and it is worth investing in a length of

14. A rare opportunity seized. When engineering work was carried out downstream the level was lowered along this fishery, providing the camera with an opportunity to record the bed, the banks, and general conditions underwater.

plastic-coated wire netting which will outlast two or three applications of the ordinary type.

You can, in addition, use willow laths (laid in two ways) in the hope that they will grow. This is used widely by Water Authorities but does cause problems, such as growth getting out of hand. What happens when the growth fails is that you end up with thousands of willow laths lying around all over the place, as well as submerged stumps. Nevertheless, it can be very effective as a quite long-term erosion stopper when it does work.

Deposition occurs on the other, convex side of the bend where water slackens its speed, allowing silt that is carried along to drop. In theory one should be able to remove it by altering the flow and so scour the stuff away. Practically, however, it is nigh-on impossible to move the flow straight across the stream to make this happen. The only cure is to remove the silt with muscle power, allowing a flow of water to ride through the area again. From choice we make this a summer job, stripping and getting into the water to really tackle the work. This particular problem of deposition leads naturally to the next problem, that of:

SILTING

Here we are considering a large area that has silted and practically stopped the flow of water. Complete dredging is usually out of the question; remember we are considering a fair stretch, not just a little deposition on a bend. The natural reaction is to call in a mechanical digger providing, of course, one could afford it. In most cases this only provides a temporary cure, because the nice deep stretch that the excavator leaves (so beloved of many Water Authorities) will have little flow, and in less than a few years the whole area will be silted again, and very shallow.

Our experience has been that most silting is caused by an excessive weed growth, more especially of marginal weeds such as reeds, rushes and kindred species. A drastic thinning here will help, plus some weed cutting (see Chapter 6) and an inducement to movement through the length by means of a small channel hand-dredged with buckets, where this is possible. An alternative would be to consider the erection of a small dam on a temporary basis upstream of the afflicted area.

Odd patches of silt that appear from time to time, for no apparent reason, are often the result of an underwater obstruction nearby. Find

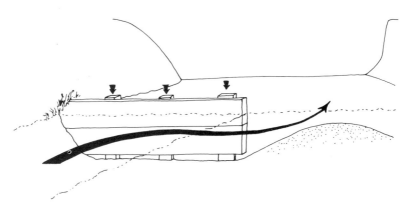

Fig. 6   Silt boards can be used to deflect the flow, to erode silt beds, or at fullstream width to create a silt trap.

15. (Left) Weed choke. Current is being directed against the right-hand bank, causing erosion. The weed on the left is resulting in a massive silt bed.

16. (Right) The same bend after treatment. The current is passing away from the bank and will scour out its own course without further danger.

that, remove it and you will have a cure. To save the muscle power required to physically shift the silt, try some silt boards. These are pieces of wide timber staked out against the current, so diverting the water into the silt, cutting it away. Our Figure 6 shows this simple item in use. A word of warning though – silt washed away must go somewhere else downstream. Check that it is not going to make more work further down the fishery!

Silting is usually a major problem in still water management. Years of leaves falling into a lake or pond, and feeder streams that have carried tons of fine silt into a resting place in the main water are the usual culprits. Quite a lot of silt can be bucketed out by hand and distributed, well away from the edges, on an odd piece of ground. Even raking the margins at regular intervals does help.

It is a stinking, messy job and one that requires bathing trunks and an old pair of gym shoes – clothing of any sort takes the smell of silt for days after the exercise is finished, and rest assured your best friends *will* tell you! But it is amazing just what quantities one can shift, especially with two pairs of hands, one using a bucket, the other man-handling a wheelbarrow. An added bonus from this sort of work is that you really do get a chance to look for insect life and its distribution when your head is constantly stuck over a bucket!

SILT TRAPS

It is no use dredging silt from a water where the stuff is still being carried into it through the medium of feeder streams. Only one cure exists for this problem and that is the construction of silt traps. These are mini-weirs, staked across each stream a short distance above the junction with the main water. They should be well dug-in at the ends and sunk at least a foot into the stream bed.

Corrugated iron or asbestos sheeting is an ideal medium and this should be reinforced at the top (above the water level) by a substantial baulk of wood. Once in place the hard work begins, excavating a large pool backwards (upstream) from the construction. It is in this area of static water that silt will drop and naturally it will have to be shovelled out once, and probably twice a year.

It sounds like a lot of hard work, but it isn't really. The structure can be quite small and the settling pool quite large; the reward in trapped

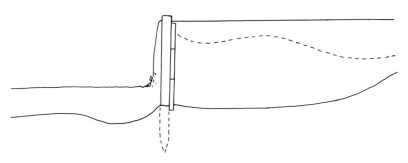

Fig. 7   Profile of silt boards used as a silt trap. The silt build-up (dashed line at right) is not always as shown but may be more to the head of the pool in its main bulk.

17. Sensible weed control on a lowland stream. Beds of reedmace are left to provide cover, breeding places and a place for insects to establish on the fishery, but are not allowed to choke and throw the current against the banks.

silt and subsequent improvement to the main water, enormous.
Naturally, these silt traps can provide protection to small drains and
streams on rivers, etc. Ken describes one, and some problems arising
from its construction (his first attempt) in Chapter 7, whilst Barrie has
prepared Figure 7 to illustrate the procedure.

SUBSIDENCE

The smug thing would be to say that this problem should never be
allowed to occur. In practice it rarely does, but we both, on occasion,
have had banks subside from causes other than normal erosion and
undercutting. This is usually a problem of lowland rivers and streams,
where the flow of water over the years has eroded soil and left high
banks of soft earth.

Sooner or later frost will split out a section of the bank – usually at
some height above the water level, and a whole section will slide into
the river below. Yet – where it can be anticipated – prevention is easier
than the physical effort of curing the result. Where banks are steep and
tend to be soft they are best sheeted over with wire netting, firmly
secured with stakes on all sides. The other cure, which should be
combined with the wire netting treatment, is to sow grass, weed seeds,

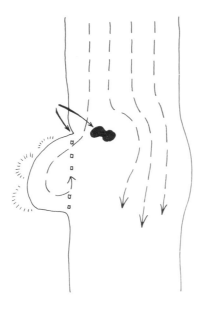

Fig. 8   Plan of erosion bay caused by
current deflection into a silt bank by an
obstruction such as a boulder or an old
log. The length between the arrow
points should be filled with netting and
rocks, and the line of willow stakes or
two-by-two (shown as squares) should
be   reinforced   with   hardboard   and
backfilled.

18. (Left) A scourhole. This one has been caused by a log jammed crosswise on the bed. (It can be seen by the broken water against the outside arm of the hole.) Professional attention, probably with an earth shifting machine, will be needed if the damage is to be corrected.

19. (Right) Erosion. Unless some prompt action is taken here, a considerable amount of land will be lost to both landowner and fisherman.

etc. on the slope so that the roots will mat and to some extent prevent further soil slip.

We know of some amusing attempts at bank saving by Water Authorities, who do at times seem to have more money than sense. The first mistake they make, very commonly in the fens for example, is to make a clay slope (usually of Kimmeridge Clay) much too steep. It often slips in classic arcuate fashion, and to prevent this they have not only used vast quantities of wire and fibreglass, but also sheets of tarred roofing felt and even polythene. One fen drain we know is actually *lined* with polythene, into which anglers regularly stick their rod rests! The use of roofing felt is a beauty: every time some unsuspecting angler stands on it he slides gracefully into the water.

SCOURHOLES
Another problem caused by an obstruction of some sort, often underwater and unseen – that directs the stream against a small section of bank, so eroding a hole out of it. Photo 18 shows a large-scale effort.

The remedy is to remove the cause, often no easy job in itself, then seal and reinforce the gap. Stakes driven into the bed and reinforced with horizontal poles can be one method. Another is to resort to heavy galvanised wire mesh – chain-link stuff – supported on posts. Once these are in place across the hole it can be filled with hard-core and a little clay (see Figure 8).

61

20. Erosion and deposition combined. Too late for any remedial action now, eventually the stream will cut through the neck of land in the centre of the picture, forming a horse-shoe lake.

Some of the scourholes can reach alarming proportions. Some, though small, herald an alteration in the course of the stream, and in either case some pretty effective action must be taken. This is one of those occasions when an approach to your Water Authority will bring more than a little help for they, of course, have a legal and vested interest in maintaining waterways.

CATTLE DRINKS

Great play has been made in books that deal with fishing, as well as water maintenance, on this quite natural area of a water. Books about angling dwell on the possible big 'uns that may lurk around the cattle drink, waiting for insects stirred up by the feet of beasts as they move into the water. Passages in maintenance books point out that the trodden banks can produce a weakness and a source of silting below the area.

We must be honest. Cattle drinks on our waters have never caused the slightest problem, nor, come to think of it, have they been the scene for a record-breaking fish. Cattle have an uncanny knack of choosing a safe place at which to drink, often on the only gravel patch for miles along the banks.

Troubles start when you shore up banks where cattle drink. They are no lightweights and quickly damage stakes laid or pegged into the ground. They also run the risk of damaging their hooves on badly sited material. We accept the cattle drink as a natural aspect of a fishery landscape and leave well alone. Rest assured that the price of livestock, being what it is, will ensure that the owner of the land will keep an eye on the scene and do something if the area becomes dangerous.

BOARDS

We have mentioned scour boards earlier, but there are two other types we should say a few words about. The first is notice boards. Take it from us, they are a waste of time and money. Nobody takes any notice of them; the empty threat of prosecution for trespassing fools nobody, and they become a target for shotguns, air rifles and general vandalism. Better by far to use the timber that would have gone into their construction for insect boards.

These odd pieces of planking, arrow-shaped where they are to be employed in running water, can be anchored out to provide a landing strip for insects on the top side, and a home for insect life in general on

21. (Left) Prevention is better than cure. This work by the Welsh Water Authority on a small tributary of the River Towy shows how trees have been laid across the length of the bank to prevent erosion during harsh winter flooding. The trunks are secured to the bank by wire.

22. (Right) Floodwater has already started a scourhole in the bank. The small opening cut into the earth has been plugged by thick brushwood, well secured. It will settle, and eventually cause a part-healing of the hole. Work by the Welsh Water Authority.

23. Laurie Manns hunting large rudd, in excess of 3 lbs, in an overgrown ditch.

the underside. Keep them in natural colours – painting, etc. is asking for them to be used for target practice by stone-throwers – and make sure that there is a fair length of cord between the attachment to board and anchor. This will allow for fluctuation in water height and throughout the year.

TREES, BUSHES AND BANKSIDE VEGETATION

Exhaust your brain before you flex your muscles in this section of fishery management. This is the Big Boys' League, and hasty action or mistakes here can set a fishery back years, because that is how long it takes a tree to re-grow. You see, we are back at the oxygen-producing game once more. Of course, what grows on the side of the bank does not put the stuff into the water, but too much bankside vegetation, or too little for that matter, can lose you the vital element.

Too much will shut out light, and this will reduce the quantity and quality of weed growth, damaging oxygen production. Leaf fall will also increase problems of silting.

Drastic removal of bankgrowth will reduce safety cover for the fish, allow too much sun to reach the water and allow it to 'heat up', encouraging prolific weed growth which can only, in time, help the water to silt solidly. Further, the removal of large trees and bushes with an established rootgrowth can help produce subsidence and remove cover for good fish under the bank.

The happy balance is to have enough high cover along or just back from the banks to provide shelter, support a normal weed growth in the water before it, together with shade and a little food from insects that will fall from branches and leaves.

24. A 10 lbs plus mirror carp to floating bait from a tiny, heavily weeded pond.

25. If you cut timber along the banks, move it well back from the edges and stack it properly. Left near the bank, the next flood (or trespassers!) will 'attract' it back into the water.

The time of year when any thinning-out will be undertaken is of importance. Avoid the summer months for cutting, which can only encourage an increased growth-rate so that the job will need to be done again, often within a matter of a few weeks. Winter is a better time and look for old, stunted or diseased wood which should be removed before any other. Cut sparingly, and walk back to look at your efforts at regular intervals. Remember that you cannot put back what you have just removed but you can always take two or three more gentle bites at the cherry.

Take surplus timber well back from the bank, sort through and select lengths that could be of use on the fishery at a future date and store them out of sight to season. Burn the rest immediately, preferably at a central site, at intervals, and away from the fishery. Heaps of cuttings left beside the bank will either be swept into the water with the next flood, or thrown back in by the next trespasser on the bank.

26. Here, a backfill of gravel against camp sheathing was necessary to provide access along a narrow bank.

Trimming and thinning of overhanging bushes to allow fishing should be your least consideration. A heavy campaign along those lines advertises the fishing and that is always something better not done. Far better is to tie back the branches that get in the way with that green nylon twine we mentioned earlier, training the growth and so minimising later pruning.

Those bare areas where cover is needed? It is a question of setting-to and planting your own. The traditional way is to cut willow slips about 1 inch thick and 3 feet long during February and March. All shoots are removed from the base to within 18 inches from the tip, and that bottom 18 inches is then pushed into the ground and firmed in place.

Those slips will grow away – willow grows at an enormous rate – but the result might not be to your liking. A mature willow placed on a bank will 'drink' 10 gallons of water daily. Multiply that a few times and one is beginning to lose water from a fishery – which might already

27. Backfill against rough but firm 'camp sheathing' of logs. Note that the high back has been retained and the rushy, boggy water margin is untouched. Such margins provide cover, spawning facilities and food.

be under pressure from abstraction – at an alarming rate.

There are trees other than willow and it is worth a few words with a local nurseryman who will recommend species suited to the local soil, and more especially the area in which you are going to plant. A small caveat on tree planting: believe it or not, planning permission is required should it take place within 30 feet of the water. Frequently, the Water Authority can pave the way for you and arrange an agreement. We are unsure if this applies on private stillwaters.

PATHS, HEDGES AND FENCES
Usually the responsibility of the owner, but most anglers, for their own benefit and safety, are more than prepared to assist with this task. We use that word safety advisedly.

Ken has a fishing pal – Glyn – who casts a nifty fly and trots a useful worm on a small stream not a million miles from Llanddewi-brefi, in mid-Wales. One night he stopped late, when he was well up in the mountains and literally miles from anywhere. It became so dark, and he was so surrounded by forestry trees, each an identical copy of the one next to it, that it eventually took him two hours before he even found the track down. With over twenty-five years police service, Glyn still reckons this to be his most frightening experience.

If you are going to fish at night out in the wilds, make sure that there is a system of white marks displayed at eye level to assist you make your way back. No problem when the water and the exit are side by side, but this does not follow in every instance.

Barbed wire is a curse to the angler, but a necessity to the farmer. It only takes minutes to cover the top strand at a point where one regularly crosses with a wrapping of plastic, or a length of old plastic hose-pipe split lengthwise and eased over the barbs.

RUBBISH DISPOSAL

Rubbish attracts rubbish. Remove it immediately and bury or burn it so that it cannot be heaved back into the fishery and perpetuate the eye-sore.

CHAPTER 6

WEED CONTROL

We implied in Chapter 3 that weed could get out of hand. Most commonly you see this in feeder ditches with emergent marginals, but it does happen in open water with soft, submerged weeds. The weed in itself is rarely any harm to the general ecology and well-being of fish, but from a manager's point of view it can interfere with reasonable fishing. So it may be necessary to remove weed. The ways one can go about this are as follows:

a cutting

b dragging and dredging

c use of weed killers, selective, contact and otherwise

d use of plastic sheeting

e use of fertilizer.

We have already expressed some doubts about (e). It's not that the general productivity of the water is not increased, but that the number of elements in the food chain is shortened: algae → limited crustaceans/insects → good total tonnage of fish. We feel that though fish quickly reach a good average size, and in fish farming on the Continent would be ready for the table quickly, the very best weights (specimen fish) may not be reached. Angling can be good for a while as the water colours and the soft submersible weeds die, but in stress situations the main product is the more vulnerable, as in all farming as opposed to natural practice. Think of all the best big fish waters. They are almost all gin-clear lochs, loughs, pools and rivers. So fertilization we would not recommend in general as a weed remover, effective though it is in this last respect.

For not dissimilar reasons, we are violently opposed to the use of grass carp. Indeed, in recommending grass carp to the nation the authorities seem to have taken leave of their senses. With the exception of water-lilies, which in part survive, and hardy emergent marginals, grass carp destroy almost everything in a small still water environment. And with the total removal of soft weed go all the small creatures that make fish healthy and fat. In the long term it is not

28. The mud and gravel hand-operated dredge in action. The effectiveness is obvious; almost a quarter of a cubic metre is removed here in one swoop.

29. (Left) The result of farm fertilizers. This small Welsh stream, normally devoid of weed in any shape or form other than a little fern moss, now 'boasts' this growth as the result of leached nitrogen and phosphates. At the moment it is a bonus; next year the build-up may be such that the stretch is covered with green slime.

30. (Right) Not such an uninteresting photograph as may first appear, for this was once part of a canal. It has now silted, overgrown with rushes, and become unrecognisable as a watercourse of any sort. It is indicative of the neglect to which we make frequent reference.

known whether they will breed in our climate (in some warmed waters they almost certainly will) so that to suggest so many carp per acre is really rough calculation to put it mildly. In fact, as we say elsewhere in this volume, the grass carp is really here to assist drainage in this country; drainage which has gone calamitously wrong. Minimum acceptable flow has come to mean minimum usual flow, maximum silting and maximum weed growth. Physical and chemical removal of weed is expensive in the water *industry*, so let's have the grass carp do it for us! Talk about putting the cart before the horse! They should mend the wheels on the cart first and, perhaps, they would discover that it would move easily.

Cutting is seen in various forms from the use of a simple weed cutter by the angler cutting a hole in the weed, to the £20,000 aquatic cutters used on drains in East Anglia. It is efficient in that weed is not destroyed. In the past it was particularly efficient on small trout streams because you could let the debris float away to the next riparian owner downstream! This is frowned upon today and rightly so, yet it still happens when the downstream club happens to be a coarse fishing outfit! From our viewpoint, however, there is no problem, for small streams can easily be deweeded, selectively, by wading and scything

72

31. The green peril. Duckweed during a hot summer has choked this section of river. Note the poor growth of Arrowhead weed in the foreground and claw tracks from a moorhen running along the watercourse – a good indication of the thickness of growth.

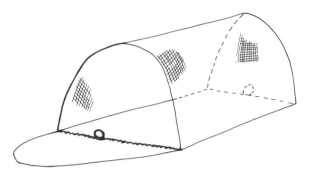

Fig. 9 Home-made mud/gravel dredge. The upper part is steel or aluminium heavy gauge mesh, and the base-plate cast iron or steel. Dog clips and ropes are attached to the rings fore and aft.

and then catching the debris downstream with a shallow net, or sacking, stretched across the stream. Whether in ponds or streams it is always advisable to let the weed rot in the margins or only just pulled up on dry land, since this allows its inhabitants to crawl back into the water. The cutting of weed close to the bottom is intended to leave in the roots, and the intention is not, therefore, to remove the weed bed altogether.

To achieve the latter you have to drag (anglers) or dredge (managers). Figure 9 illustrates a home-made yet effective dredge that

32. (Left) The 1500 gauge black polythene (invisible now) has eliminated water-lily growth down the middle of this carp pond.

33. (Right) Divers dealing with a polythene sheet that has not been properly weighted. Note that old scaffold poles make good weights and that winter is the time to lay polythene, before the weedgrowth begins.

34. (Left) One of the most dramatic (and expensive) weed cutters in the business, seen here decapitating weed on a small shallow drain.

35. (Right) Still a good pair: weed cutter and rake. The double-edged rake is a little on the heavy side. The cutter blade is of soft steel and can be sharpened easily with a file.

can be used on small lakes and streams. The effect is to really grub out the roots and the mud/silt itself so that you have a few seasons respite from that particular weed bed. Anglers' drags (Photo 35) are also very effective but only partially remove the roots, simultaneously ploughing the bottom. The last effect usually sweetens the bottom sediment by releasing methane, but also allows for vigorous regrowth in a season or so. Of course, anglers on many waters keep the swims regularly raked in this way and if they are prepared to do it – and the resulting fishing is worthwhile – it saves the management the problem of arranging for extensive dredging. Partly for these reasons we suggested earlier that even small waters should have deeps in excess of ten feet, preferably fifteen feet. This discourages weed growth in some areas at least and lessens the dangers of winterkill.

The use of weedkillers in water was regarded with suspicion for years by anglers but our discussions with the Water Authorities, as well as some personal experience, convinces us that in fishery management terms it is not really all that expensive (say £40 an acre) and can be very efficient in that it is not a final solution. Selective weedkillers are available that kill only those weeds to which you physically apply them. And some do kill only certain kinds of weed. By judicious use of boat, spray and/or polythene pipe, it is possible to thin weeded areas very effectively. None of these weedkillers is toxic to fish and invertebrates, and in biochemical terms they degrade not long after effective use.

It is clear that we are in favour of *a–c* above and anti-*e* or anything else that provides a 'final solution', since it may do the same for your

75

fish life. On individual waters there are always little tricks. For example, a canopy of overhanging branches discourages some types of weed – but the tree sheds leaves! In some cases removal of a row of trees can reduce weed growth by opening up a water to wind action, which also oxygenates.

Finally, there is the useful polythene sheeting method and this we have left until last because it is in rather a special category: after all, if one fishes over it one is fishing in a polythene bowl so to speak!

CONTROL BY PLASTIC SHEETING

Cutting, raking and digging are the three most usual approaches to weed removal. There is another, more subtle but very, very effective method which is suffocation, by means of plastic sheeting. The stuff to

36. (Left) At least half of this pond is overgrown with lesser reedmace.

37. (Right) A once-deep fenland lode now choked with vegetation and rubbish – *including* the kitchen sink!

38. How *not* to plant lilies – namely, in a water that is too shallow. This nice little carp water needs only soft weeds; hard weeds will overgrow it inside a decade.

use is 1500 gauge sheet plastic in the colour of your choice, although it will be left permanently in place and eventually will cover with mud, silt, etc. and become invisible where it lies. It is not a cheap product to buy, but bearing in mind you are getting a permanent job the cost averaged over several years of toil and sweat by conventional methods of control is not too bad.

In theory, the edge of the sheet is slid into the water and drawn across the bed from a roll. In practice, air bubbles get under the sheet no matter how you try to feed it against the bottom, eventually causing a bulge that leads to an enormous bubble of plastic poking above the surface. Barrie has recently 'done' one of the College lakes at Cambridge with this method, and the problem occurred with him. The answer is to secure a steel pole of some sort (scaffolding is ideal) the same length as the width of plastic sheet – 10 feet is the usual – along the leading edge of the sheet. To this pole a rope is secured at either of its ends.

39. A tributary of a Northumbrian trout stream, dredged and canalized as part of a drainage scheme. The natural lush bankside growth is apparent; equally rich growth normally occurs in the water itself.

Both ropes are passed across the water and a straight, slow pull then takes the weighted end sliding over the bed to the other side, with the plastic following and lying flat. It works well on small stretches of water, but large areas and lengths of stream or small river need the plastic to be weighted at regular intervals with long weights – bricks and stones are not good enough – otherwise the sheeting will lift and either disappear downstream, or rise to the surface.

Barrie's solution to the problem on the College lake (quite a fair stretch) was to call on the service of some underwater divers who put the sheeting to bed and then weighted it with scaffolding poles at regular intervals. The job was far easier with help from the divers than by any amount of trial and error by people keeping two feet on dry land.

The success of this method can be judged by Photo 32, which shows the swath of weed-clear water stretching away, lilies to the right and left, but nothing where plastic has been laid down the centre. Moreover, the plastic is not visible, being black in colour and quickly covered in algal slime.

On a miniature scale small corners, odd swims, weedbeds with heavy root growth, can all be tackled by the aid of plastic bags, fertilizer or dustbin liner types, opened along two sides and spread out, then weighted with bricks. Naturally, you will have to be in fairly shallow water for this type of wading work to be done.

As well as providing complete control the method will also enable controlled weedbeds to be planted and maintained. Lay the sheeting on the bottom, cut a small hole through it and then plant your weed through it. Roots may well spread along under the sheeting, but it is impossible for shoots to appear and spread upwards to the surface, other then where the initial planting was made. A worthwhile consideration, especially where lilies are contemplated on the fishery.

By and large, black is the best colour because it cuts out light and prevents growth beneath the sheet, eventually killing all plant life below it. For water-lily beds the sheet needs to be down for eighteen months to achieve complete removal of the root system. It is possible to fish over the sheets, but one does catch the edges occasionally, and the weights. Having taken out the sheeting it is certainly several years before lilies gain a foothold, though soft weeds may be back in two years.

CHAPTER 7

CASE HISTORIES

So far we have discussed a considerable amount of theory and now, perhaps, is the time we should show that we have used – and misused – some of this knowledge ourselves on the various small waters that we have managed over the years. We discuss them and the management problems that have centred around each with no attempt to disguise our faults, in the hope that our sometimes obvious mistakes may help others.

Ken now describes Leg o' Mutton, a small lake and a stretch of a small South of England lowland river which he has called Oldways. Of necessity, these are imaginary names used to protect both fishery and owner alike.

LEG O' MUTTON

Despite our cautionary words on obtaining a water in Chapter 4, Ken was presented with this lake on a plate. The owner, a shrewd and hardworking business man, had been plagued by poachers and vandals to such an extent that he offered the water to a local angling club on the understanding that they would bailiff it. Unfortunately for the club, but fortunately for Ken (who lived close by) they decided their membership was too large for such a small fishery to be fairly shared so the club secretary mentioned Ken's name.

The path to the water was wide enough for a vehicle to travel along, albeit with care, and lay through a fine old stable yard, across a couple of meadows then down into a small deciduous wood underplanted with rhododendron and bramble. The wood actually lined one side of the lake and hid a large part of it as one approached, so that it took several minutes on the first visit to work out its shape which, of course, has led to Ken's name for the water. (See Figure 11A on page 90).

The acre or more had been formed by putting down a dam across a stream at the north end; a thick clay bank which had been improved by several ash trees, the roots of which had helped bind the soil into a solid mass. There was a concrete overflow in the north-east corner, excess water falling back into the interrupted course of the stream and away to heaven-knows-where.

The main feeder stream led in from the southern end, a typical lowland effort that had drained surrounding meadows and naturally was inclined to carry a fair amount of silt. This had led to a bog area being formed around its junction with the lake, and there were a dozen or so hawthorn and elderberry bushes together with a lining of greater reedmace some four feet wide at that point, before the bog dropped away into deeper water.

A second feeder ran down through the wood and joined the east side. Although it was mainly a flash-flood effort and contributed nothing greatly in the way of flow during summer, this feeder ultimately presented the greater problems of the pair. Dropping rapidly downhill between the trees it had scoured a deep course through fine gravel and sand, and with no grass or weed growing on the banks to bind the soil particles an enormous push of silt had taken place over the years, settling in the shape of a wide bar that protruded into the lake for a distance of twenty feet or more.

There were further problems along the east side. In the course of time a line of stunted willow trees had grown, fallen into the water, pushing out adventitious roots into the shallows and re-growing. There was an Amazon jungle of live and dead wood extending outwards and stretching along half the length of the bank. But by contrast the west bank was bordered by a meadow, its edges lined with greater reedmace to a width of six feet, totally unfishable, but allowing necessary light to spread across the surface of the lake.

There was not trace of any sort of water weed other than the reed mace already mentioned, whilst the lake bed appeared to be of leaf mould on clay and gravel. Once we had launched a boat we discovered that the depth ran uniformly, south to north, from one foot to a maximum of six feet. (Note that at this point the Royal 'We' has suddenly appeared in the narrative). Ken spent a week carefully mapping and watching all there was to see before deciding it was all too much for one man, so he promptly called in Ted, Roy and Bob to form a syndicate. All were keen, lived fairly close, and were as impecunious as Ken.

We studied the map and the water itself for a further three months, during which time we discovered that the water was heaving – literally alive – with small common carp. The biggest landed was $1\frac{1}{2}$ pounds, which weight represented a minor miracle, for there appeared to be little or no natural food to support the head of fish, at least not in quantities sufficient for the enormous population to thrive.

By sieving, pulling reed mace and examining the roots and soil, we discovered caddis larvae, leeches, a few pond skaters, dragonfly larvae and possibly the biggest food source, midge larvae. The one odd great diving beetle which flew into Ted's face at fearsome velocity late one night, nearly rendering him senseless, was not included in the list.

Through the following months problems and possible cures were discussed, judged and where possible, acted upon. Some tasks could be undertaken straightaway and assessing the pH value of the water was just one of them. We had a test performed through a mutual friend at a university (for free) and the result was a value of 6.3, which was on the acid side and only, perhaps, to be expected in view of the rotting leaves and branches across the bottom of the lake, plus the stream of water that fed intermittent but often copious supplies through the barren wood.

It could be better – but attempts to cure the situation were temporarily set aside whilst we grappled with the fallen willow trees and silt which had collected by the hundredweight around and under the roots and splayed branches. We agreed that it was the smelliest job we had ever undertaken in our lives. Waders and old clothes soon stank of the black silt, but gradually the branches were removed and burnt, to be followed by barrow loads of silt that was taken back into the wood and scattered so as not to arouse unwanted interest in our activities. There was just one highlight during the month-long task – when Ken fell backwards full length into the water. It was unanimously agreed that the single word he uttered adequately described the smell that he exuded afterwards.

The south end, where the main stream entered, was cleared of debris and small trees and bushes were trimmed, but not removed. It was decided that their root systems, together with the lining of greater reed mace at the water's edge, was an asset, filtering out much of the silt as it was carried along. A channel was cleared and the reed mace thinned somewhat to prevent it extending further into the main body of the lake.

We would have preferred to erect a dam somewhere along this stream, but were prevented by the wide and shallow nature of the stream's course. Nowhere were the banks higher than a foot, and it was impossible to fasten the timber sides and supports of the structure into them. Try as we would, water would flow round the ends, flooding quite a wide area and adding to the problem of silt dispersal rather

40. Stream clearance. This feeder (at Leg o' Mutton) is being opened to allow light and air in. Branches that have been cut are laid along the course to build up a bank, where excavation with buckets can be carried out with relative ease.

than confining it. Eventually we decided to plant suitable weed beds beyond the reed mace, which we hoped would act as a further filter, trapping silt in shallow water which we could easily dredge.

However, the woodland feeder on the east side was in an ideal position for a silt trap and this was erected from wood and corrugated asbestos. The wood, a thick branch, was set between the highest points of the banks some ten yards back from the junction with the lake. Against this, set well into the bed and sides was a sheet of corrugated asbestos with a 'V' overflow section cut into the top centre, allowing water to draw away from the edges. Following the installation, an afternoon was spent digging behind (upstream) of the trap which produced an enormous hole, the settling pool for silt which would fall whilst the water was relatively still. Earth that was removed was used to reinforce banks and the superstructure of the trap, rendering it completely watertight (see Photo 40). Each year the pit has been dug out again, accompanied by beer drinking and rude story telling to ward off boredom.

Below the trap where water falling over the lip would land, some rubble would have to be provided to prevent erosion of the bed and it was here that we had a brainwave to improve the pH value. If, instead of introducing rocks and bricks, we used large lumps of chalk we would not only prevent erosion but possibly help to kill the acid nature of the water. Of course, the bed of chalk would have to be fairly large and several hundredweights would be needed, but the problem was not insurmountable. Away to the north of the lake they were cutting a new motorway through chalk outcrops, and an approach to the Irish foreman and an explanation of our requirements produced the statement that we could help ourselves but, if the road collapsed, we would have to bring it all back! We think he was laughing. Three bootloads by each of two cars soon produced the required amount and to date no request for its repatriation has been received.

Another silt-clearing struggle demolished the bar that spread out from the junction of this stream with the lake and here we formed an island facing the outlet itself, reinforcing it with branches and clay until the current entered and turned sideways, alongside the lake, instead of charging straight out towards its centre. This effort ensured that silt which did get carried over the trap would settle where it could easily be reached, and saved our having to wade and struggle with endless buckets away from the safety of dry land.

By the time these tasks had been accomplished a year had passed. It was late autumn and high time we did something about the fish population. There was no doubt that we were overstocked and not only

41. (Left) Where a feeder stream carries silt into a lake it is advisable to direct its course sideways along the bank, where excavation with buckets can be carried out with relative ease.

42. (Right) Planting marginal weed. In this case a box of bricks is being made to support loam into which plant roots will be spread.

were a majority of the carp stunted but we had also caught two with dropsy, showing that the health factor was not all that it could have been (see Photo 49).

It was at the time of River Boards and regulations on netting and stocking were not as severe as they now are under Water Authorities; in any case, our little lake fell just outside the jurisdiction of two Boards – a section of no-man's land that allowed considerable licence. After a long council of war, we decided to contact Rags Locke in Hampshire, probably one of the best known fish farmers in the country.

Rags arrived on a bitter November day complete with Land Rover, water tanks, oxygen aerating equipment and nets. He made an immediate impression on Ken by lighting a fire with six wet green sticks and one match; after that anything could have happened but, unfortunately, didn't. The carp were obviously well down on the bottom in the mud and despite a freezing day hauling on ropes, we finished with Rags taking away a paltry two or three hundred undersized fish. But revenge was planned.

We set the score straight during the following May when Rags came down with all his previous paraphernalia plus a small boat and an electro fishing set. We also produced a small fibreglass dinghy, and both boats were soon lashed together and the electro gear secured aboard. Rags' experience told him that many small fish would be in the reed mace round the edges and under tree roots at the north end of

43. One for the insect count. Eggs of a watersnail show that stock introduced the previous year on this water have established themselves.

the lake, and it was for this reason that he had included the electro stuff which normally was reserved for shallow, running water. His hunch paid off and we netted several hundred tiddlers that were stunned, all in a position where they could never have been reached by the net or other means.

From there on it was the net; slow work passing a line around sections of the lake, harder work hauling. Take our advice – when first you go netting, go with an expert. Brute force and ignorance are useless for the job and speed, combined with an ability to haul on both head and foot ropes evenly and smoothly to prevent trapped fish escaping under, or jumping over the mesh requires discipline – and a lot of practice.

Our hauls were good and fish recovered were quickly sorted, those under the legal limit and even a little above being gently placed in buckets which Rags then transferred into the aerated tanks aboard the Land Rover. The larger fish were returned, and we carefully sorted through the debris dragged into the net by the footrope for signs of further insect and mollusc life without, unfortunately, any increase to the list of those we had already noted.

44. Professional equipment. A common sight at Leg o' Mutton, with tanks for carrying fish, nets, buckets etc.

It was impossible to count the numbers of fish that were taken, but we had, by the end of the day, measured thirty-five full buckets of fish and turned them into the tanks which were bulging with carp in the prime of health; in fact, we only had to destroy five diseased fish, two with fungus and a further three with dropsy.

Before leaving, Rags offered sound advice on the future of the stocking programme. He reckoned that the breeding rate, influenced to some extent by the size of water and especially its ability to 'heat-up' quickly in the warm months, would necessitate our having it netted every two years or so if we wanted to build a stock of larger fish. Alternatively, we could make a vast improvement in the food store, but that would take time and there could be no guarantee that it would succeed in reducing undersized fish.

We thought carefully about his words. The arrangement was that Rags provided the netting service in return for the fish, which left us with no pecuniary liability. But the thought of repeated netting, especially over the weedbeds that we hoped to establish, did not appeal. We would adopt other tactics.

The next few months were preoccupied with weed. We decided that the best source for stock would be the nearest stretch of water, working on the assumption that imported weed could import disease, and disease from water close-by would probably have been introduced already into our lake by wildfowl, of which we had a small stock in the shape of moorhen, coot, duck and a solitary heron. A small pond just half a mile away supported several fine beds of starwort, and we took a fair stock and planted this carefully, taking care to spread the bulk into small clumps and planting them by the inlet stream at the south end.

Next day it was gone. Not a trace. So we replanted with the same result. Perhaps, we thought, the carp were having a field day and grazing off our efforts straightaway; if so, we would have to provide protection and this was achieved by the means of a wire netting cage (Figure 10) set over each clump. Within a few days this weed also disappeared and we then decided to do what should have been undertaken in the first instance; set watch.

Two evening later we saw our latest planting chewed away by the moorhens who fed as if famished. Apparently they could push their heads through the mesh and feed with ease – but when we examined

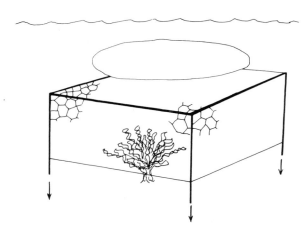

Fig. 10   Wire netting protective cage, weighted on top with a rock, to cover newly planted valuable plants. The prongs rest upon and sink into the bottom sediment.

45. (Left) Setting up and checking electro fishing gear. The two boats were secured side by side to provide a stable platform.

46. (Right) Electro fishing the shallows, near a bank. This method of taking fish is seldom effective in deep water, and provides best results on streams.

the site later we found that they had only taken the foliage, not the roots. We decided to leave well alone and let nature take its course.

The speed at which carp could reproduce was instrumental to our making Leg o' Mutton a dual-purpose water. After considerable discussion we decided to introduce pike – and in autumn we secured four specimens in the four- to six-pound weight bracket, all of which had been lightly hooked. They were transferred from a gravel pit six miles away by means of a wet sack and travelled well. Let us hasten to add that our action at that time was entirely legal and there was no restrictive legislation on stocking as there is today.

Strangely, we could reach no final decision on how many pike we should introduce or rather, how many would be necessary to establish the species. Looking back, I feel that we showed great restraint in restricting our fish to just four, but know now that we were right. It is easier for nature to achieve a balance rather than having an imbalance forced upon it. From then on it was a matter of sit back and wait – not just to see how the pike fared, but how all the various improvements would 'jell' together.

Four years later there was a good head of starwort around the southern area and this filtered quite a fair amount of silt, which was dredged from both feeder streams, together with the silt trap, on an annual basis. The weed and reed mace was also kept under annual control and a pH check revealed a value of 6.5, no great increase but a move, at least, and in the right direction.

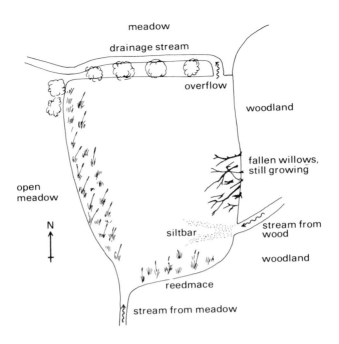

Fig. 11A   Leg o' Mutton lake before and after.

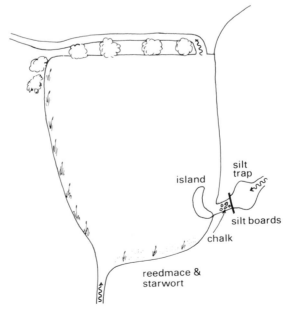

47. Hard, and cold work. Hauling a net, carefully taking up the slack on the bottom rope under which fish will escape.

Now, nine years later, the lake has possibly settled into its future norm. We have taken pike to sixteen pounds and young jack are coming on. Carp have increased in weight to include a fish of $4\frac{1}{2}$ pounds, and the quantity of small fish has decreased substantially. Certainly we prefer quality to quantity. The weed has run amok in places and with hindsight we might, perhaps, have been better to have introduced something larger, such as the broad-leaved pondweed or similar which would have been easier to cull.

Insect life is a little richer, possibly because of the intensive water snail drive we organised over two days, transferring marsh snail, great ramshorn and flat ramshorns in fair quantities. There has also been an increase in the caddis larvae population and we have recorded the water louse as being resident.

The most difficult task? One that has, so far, been unrecorded – that of dealing with poachers. One or two have been apologetic, the majority truculent when asked – in the nicest possible way – to move, and some have been violent.

48. When you have a fair assortment of the bed mixed with the fish you know that not many will have escaped. These carp were taken at Leg o' Mutton.

We have now arranged the banks so that there are just four very apparent places from which to fish. Sunk, by means of thin wire, is a scaffold pole wrapped in barbed wire that sticks across the bottom of each swim and it is amazing just how much fishing tackle we recover from these obstacles. It is, of course, a simple matter for us to lift the wire from its hiding place and to remove the obstacle when *we* want to fish, replacing it when we go. Most times, however, we are content to work for our fish, facing difficult casting places. This is something that the man who poaches seldom seems to want to do.

If there is any measure of success in the story of Leg o' Mutton, it will only serve to show the inconsistencies and shortcomings that exist within this incomplete story of Oldways that follows. By incomplete we

49. One good reason for netting – a carp with dropsy, which should be destroyed.

refer mainly to the length of time that the fishery has been under management, a scant three years. Many anglers may feel that to be a sufficient time to decide what is successful and what is not on a fishery, but you can take it from us, it isn't. We think a minimum period before judgement of any kind can be passed is five and, preferably, ten years. But to the facts.

OLDWAYS

Michael is a yeoman farmer whose roots can be traced through both his farm and farming itself for several generations. To him land is valuable, not just from a financial basis but because it is something impossible to manufacture and, therefore, too precious to be wasted. When Ken approached him and asked for permission to fish Oldways it was granted, but with it came a request for an assessment of the fishing and its possible potential. From that meeting came the chance to improve a difficult fishery, not only for the immediate sport which it could provide, but as an asset to be handed to future generations.

The water has the appearance of a typical small lowland river, varying in width from ten to eighteen feet along the three-quarter mile stretch that runs through Michael's land. The rich loam soil, drained by land drains into the river, is farmed on a hay and cereal rotation with strong emphasis on natural, as opposed to artificial fertilizers. And it is the very richness, the friability of the soil that has lead to the biggest problems on Oldways, those of subsidence and erosion.

Summer droughts and floods, together with winter's frosts, have helped to cut the river bed down through the soil until the everyday water level is often ten feet below the top of the banks, themselves level with the fields that stretch on either side. The steep 'V' to the bank sides are of soft earth and this subsides at regular intervals into the water below. To compound the problem is the fact that the natural course of the river at its lower end, towards the boundary, meanders widely, leaving sharp bends that trap the silt and this has lead to a weed problem. (See Figure 11B on page 96).

Some control to the flow of water is provided by a sluice at the upstream end of the farm which is worked on an automatic electrical system which raises the gates once a predetermined depth upstream of the structure is reached. In theory the idea is good, but in practice it is an all-or-nothing effort that provides a raging flood without warning and is probably responsible for much of the subsidence along the banks. Unfortunately, the gates seldom remain open long enough to sweep silt cleanly away downstream and off the fishery.

Both banks are lined with oak, ash and an assortment of hardwood bushes that provide ample shade and would appear to have introduced the right amount of cover across the water, encouraging both weed growth and insect life in acceptable quantities – a little too much in fact for the former. Included in the first weed count was willow herb, loosestrife, meadowsweet, reedmace, arrowhead and a small quantity of hornwort.

The insect count produced stone, caddis, dragon and damsel flies with their larvae in abundance, whilst a count taken during some weed clearing further revealed three types of water snail, water bugs, beetles, skaters, boatmen and a few, very few, leeches.

The variety and quantity of insects must have been sufficient for the water for there were plenty of fish and they were in excellent condition, although not covering a very wide range of species. Possibly the biggest head count was of chub, followed by diminishing quantities of bream and perch with a few, very few, brown trout. There was always a fair head of pike running up to the ten pound mark, beautifully marked fish which appeared to have colour etched in their sides. Naturally, with the sea just ten miles away, there were innumerable eels that made the use of a worm in any shape or form a nightmare and which attacked the biggest deadbait presented for pike with vigour and stamina.

Set out on paper and viewed in the cold light of day there would appear to be little that could be needed in the way of management and improvement on Oldways; yet, in practice, there are problems which have proved some of the toughest Ken has ever had to tackle. At the outset the most difficult task centred around the high banks which obviously could not be left to crumble away, if only because the silt build-up was encouraging a spread of the wrong type of weed further down the fishery. On first inspection, curing the subsidence appeared to be beyond the skill of an amateur and possibly Ken would have abandoned thoughts of a solution had he not waited, as Chapter 5 suggests, a clear twelve months before making decisions and starting serious work.

Observation over that period showed that sections of the banks which did crumble during winter finished up by spring in the shape of a lazy 'U', so that by the following winter a sharp 'V' was presented, and the process would start again. The thought occurred that there might, at the 'U' stage, be a chance to effect a cure by sowing grass seed which would bind the soil and Ken accordingly sent out to an agricultural seedsman and purchased fourteen pounds of a mixture which would provide a quick-growing spread of coarse grass.

As each 'U' appeared the sides were further shaped away and seed broadcast, then raked in to prevent birds from scoffing the greater part. After three months there were signs of a sprouting and spread of growth so that by the end of summer the three sections on which the experiment had been tried were well covered and appeared to have bound the soil firmly. Photo 50 shows the final effect.

During the following winter the sections held. There were two or three tiny patches where some slip took place, probably where Ken didn't study the contours sufficiently, leaving sides too steep for the grass to get a firm hold. The following summer (the third year since Ken took over the water) helped promote an even thicker growth from grass plus various weeds, including nettles and docks that have provided reinforcement. As an alternative to shoring the banks, which would have been financially as well as physically impossible, it is an unqualified success and, as each section of bank slides into the 'U' shape, treatment will commence and continue.

If you look at the map of the fishery (Figure 11B) you will see that there are two rather long, straight and shallow stretches through which water passes with some speed, even during times of low water. This, of course, has meant that the stretch is practically weedless – roots are washed out within a short time. It is an obvious area for a shallow dam, slowing the current and providing a little depth so that plants could get a hold. The idea has been put into practice with one of Ken's wooden dams (see Chapter 5), but with every flood the raising of those automatic sluice gates upstream has resulted in both dam and weed being washed clear. More permanent structures have suffered the same fate and now the project has had to be abandoned.

But further downstream, where the river meanders in a succession of

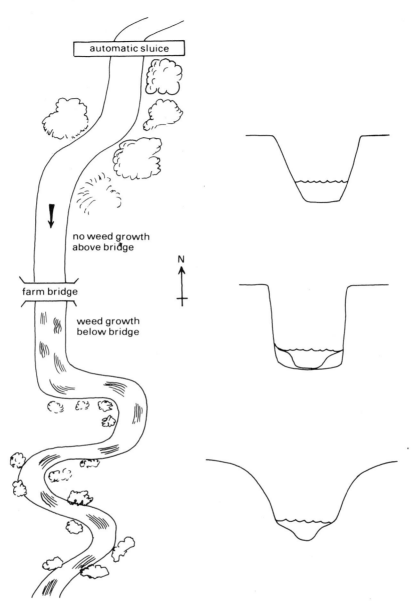

Fig. 11B Oldways. The upper cross section is roughly V-shaped and has had winter silt washed away during the summer; the central section is when winter floods have pulled silt from the sides; the lower section is ready for sowing and the steep upper parts have been raked into a wide U-shape.

tight bends, a different situation presented itself. Deposition from the eroded banks, plus the addition of silt from upstream subsidence, left shallow areas beloved by reed mace and other marginal plants that were established right across the bed in places, encouraging further silting.

A vigorous attack has been made and many of the weedbeds, together with their root systems (most important!), have been thinned back or removed altogether and, after draining to allow insects to fall back into the river, carted away and burned. Some complete beds are still in place, largely where they cannot obstruct the stream. These supply shelter, together with spawning areas. An effort will be made to enlarge on the amounts of hornwort present and possibly introduce further underwater weed in the future, with thought of spawn and its increase in mind. Removal of the weed has allowed the use of siltboards and banks of thick sludge are already moved on, or have been bucketed away from the banks. But the project will not be completed in five minutes – nor five years probably.

There was room for much improvement in the fish at Oldways, both in the spread of the various species – there was a tendency for fish to 'crowd' the bottom end of the water – and in the species themselves, which did not arouse a great deal of interest. Ken contacted the local Water Authority who said they would be delighted to send their bailiff for a chat, and he duly arrived.

He walked the banks, listened, inspected and immediately suggested the most sensible thing would be to investigate the present head of fish, after which a decision could be made on which further species, if any, could be introduced. So it was a case for the electro fishing gear and, on a spring day a few weeks later, three people were soon sweating with the equipment.

There is something of the small boy in every angler when it comes to electro fishing a river. At a glance it may appear that the water could be fish-free and not worth a second thought. But often the electrodes show a different story and as each fish floats up you begin to wonder just why it chose to live in that particular position along the bank, where most probably you would not have entertained a cast had you been using a rod.

The biggest surprise came with a head count of the chub. They were not in the specimen bracket, but plenty were in the three pound region that gave some fresh fuel to ideas on angling tactics. Knowing the voracious appetite of a single chub, the combined appetites of fish that were seen that day immediately ruled out thought of introducing more unless there was a clear-out among those present. From Ken's point of

50. Subsiding banks. Those on the left have been sloped into a 'U' as opposed to the opposite, untreated banks. Weed seeds etc., planted on the sloped left banks have held and strengthened the sides. A picture taken at Oldways.

view he would have welcomed a stocking of grayling to provide sport with fly and float for nine months of the year. An alternative species, taking into consideration the type of water, would be roach or dace, but they could well finish up second best in the food stakes when it came to competition with the native chub.

After an hour of talking a deal was done. Subject to approval at Fishery Officer level, a netting of chub would be removed and a stocking of grayling provided in their place, which would come from the northern part of the county. The exchange would take place when (and that was the operative word) it could be fitted in with the minimum of cost.

So Oldways waits now for the new fish, while Ken works on the fishery problems. Projecting into the future it is hoped that in another three years or so most of the bank problems will have been improved upon, but in all earnestness it is doubted whether they can be cured, other than with professional help. Marginal weeds are under control now that silting has been reduced and careful thought, plus consultation with the Water Authority, is going on with a view to introducing a further species of submerged water weed – one that will improve the level of insect life by providing more protection, together with better spawning facilities when the grayling become resident.

With the possibility of fly fishing there will have to be a hard look at the physical features along the banks so that a line can be cast without constant tangles. Obviously the problem of wading along some of the sunken stretches, well below the banks, must also be considered and finally there is the eel problem that must, if the grayling are going to breed, be put in perspective. Oh, and there are a few mink that appeared last year. Solve one problem and another presents itself.

The problems go on and on as you can see from Ken's experience at Oldways. Management means *continued* management, not a one-off job. Of course, all anglers become involved with club work from time to time, attending work parties and so forth. And if you are a newcomer to such events you soon realise that a hardcore of club officials not only do the organising but the hard work too: they are the 'managers'. We hope that the numerous examples throughout this book will help you, even if you are only occasionally assisting the managers rather than running a fishery yourself as a lessee or as an owner (Chapter 4).

The following case histories, also largely disguised by name changes, contain a variety of do's and don't's that Barrie has experienced in tending these waters more or less as a manager.

CYGNETS

We must explain the title. It refers to the disappearing cygnets problem which we had season after season. What problem? you may ask. Don't we want *all* cygnets to disappear? Well, no we don't. Although they are a bit of a nuisance at times, making use of floating baits on this lake quite impossible, they are, like most wildlife, nice to have around. True, they mess up the banks in places, in quite filthy style (something of which the general public is quite unaware), but still nice to have around. What happened each year was that four or five cygnets hatched but when they reached small goose size they disappeared, one by one, so that by the end of the summer only the parents were left. The adults returned the next spring and exactly the same thing happened. We discovered eventually that the local farmer was eating them! Quite a regular and excellent supply of food when you think about it, illegal though it was, and is. Things have changed in the last twenty-five years and most cygnets today stand a good chance of growing up. Since swans live to about twenty-five years, quite a high proportion of today's swans might be quite old, but since no nationwide work has been done on the subject, we don't really know.

Anyway, they weren't really the problem on Cygnets lake. A couple of years before our syndicate obtained the carp water, there had been a bad pollution and fish kill – carp, pike, eels, tench and rudd were found dead in quantity and size. It was rumoured to be a total fish kill. It was. Although we investigated very carefully and spent hours watching, and fishing, the small lake was totally devoid of fish. Tests by the Water Authority showed it to be perfectly good water, and so we got to work on it, stocking it with fast-growing carp and, at times, with anything we could get hold of because we were not aiming for a carp-only water. Our survey showed that the most likely cause of the totality of the fish kill was that the water was very heavily overgrown: silted and thickly weeded; with considerable areas and depths of very black and foetid mud; with a superabundance of overhanging and submerged willows, and with vast areas of rotting twigs, branches and leaves around the margins. Everyone likes an overgrown water but on this one it was difficult to get a rod between the trees and there were few swims as such.

So our first major task was to cut out the willows. And here's where we made our first mistake. Any fool can cut down a tree: it takes proper management to get rid of all the small branches. Our workers either left them where they fell, or they tried to burn them on the spot, and inefficiently at that. We got swims cut right enough, but all about them were unburnt and unmoved masses of dead branches (which overgrew with brambles in a season) or half burnt piles of logs. Several banks became little short of an ugly mess which no-one, eventually, was prepared to tackle.

Whoever manages the work programme should be quite ruthless in forcing through a proper programme under such circumstances. The steps needed can be summarised as follows:

1. On felling the tree, drag it away from the lake to a central site. On a rectangular lake you can have one site on each side. Bearing in mind

51. Planting ferns in a damp and shaded position.

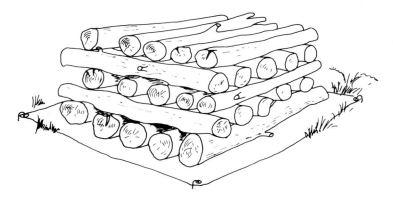

Fig. 12   The correct way to make a useful log or pole pile. Black polythene should be used as a base.

the size of waters we are dealing with in this book no member need drag a tree for more than one hundred yards.

2. If the tree is too large, cut out the trunk and any useful poles, stack them neatly at *maximum length* as shown in Figure 12 (*not* as in Figure 13, where the poles would rot and be useless). Remove the smaller branches to a central site.

3. At the central site make a bonfire, but *do not light it*. So much time is wasted by those who prefer to tend fires rather than work properly. A bonfire should only be fired at infrequent intervals and then only when it reaches November 5th proportions.

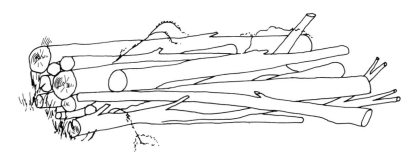

Fig. 13   The incorrect way to store logs and poles.

52. How *not* to leave a cut log. Even with an axe a much neater job can be made.

4. Even at the central sites, do not burn logs and poles, as these can be useful round the lake later on. And if, eventually, they have to be burnt they can be burnt more efficiently and usefully in someone's home. You can even sell them and obtain funds for the club.

5. The consequence is that the swim is prepared and tidy from the outset. The result of one's work can be seen and the swim can then be cosmetically treated for actual fishing. Just supposing that it is not possible to move timber to a central site (you may be on an island, for example) then....

6. Cut out the logs and poles and stack them neatly as in Figure 12. Prepare the scrub for burning on the spot. The way to do this is shown in Figure 14. It is also preferable to site the fire in a hollow, if possible, and to cut the brush into short pieces with a machete. And, again, only fire when you have enough to burn.

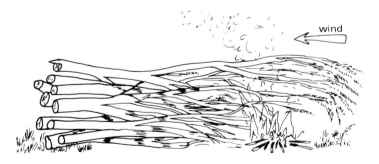

Fig. 14 The correct way to stack brushwood prior to firing. As the fire takes hold the poles are pushed into it from the left.

53. (Left) Big stock roach being introduced at night to a Cambridgeshire gravel pit.

Eventually we adopted this procedure at Cygnets and the place really began to look like a water quite quickly. The next step may seem rather minor, but is very important indeed. We refer to raking the margins. Most small, shallow lakes are destroyed from the margins outwards by an accumulation of willow twigs and branches which trap sediment, encourage brambles and eventually deaden the water to fish and other aquatic life. So at intervals a team should walk the margins raking out all rubbish.

We also designed an effective system of removing black mud (bearing in mind that we didn't have the resources to bring in diggers). This is illustrated in Figure 9, and in Photo 28. It does depend upon being able to work from both banks but if the blade is reasonably sharp it can be used to grub out unwanted soft weed beds (but not reeds and rushes). It's a filthy task and requires two or three people on each end.

The walks between swims were heavily overgrown with brambles, and the situation had been worseed by our predecessor's bad

104

54. (Right) A big roach being
stocked, not long after spawning. This is
not generally the best time, and the
travelling time must be very short.

management practice. Do *not* burn out bramble bushes unless you
intend landscaping with a bulldozer afterwards. Instead, work the big
bramble bushes you actually need – and you will need some to provide
cover, for aesthetic reasons, for brambling(!), and to leave
overhanging the water. Then trim these bushes round their margins in
order to curtail the runners, and encourage the bushes to grow high.
Now comes the hard bit. It is absolutely essential to dig out the roots of
the bushes you do not need. This is fairly straightforward unless the
bushes have been burnt down in previous years. If no burning has
gone on then cut away growth, find the main roots (which are
relatively few in number – say one every five to ten square yards) and
grub them out. If the area has been burnt you will find dozens of new
plants, all with small root systems. Fortunately they can usually be
pulled out by hand if you wear heavy canvas gloves, but is really hard
work and the removal must be thorough. An area we last cleared over
ten years ago is still almost clear of brambles.

The carp we introduced to Cygnets eventually solved all our soft weed problems by eating them. Over a period of perhaps six months they grubbed out the lot and the lake, a gravel pit, remained relatively coloured thereafter. Of the other species of fish introduced most made good headway: after all, the water had a good pH, ample food, and needed only the dead areas clearing and deepening, coupled with wind aeration after removal of the willows, to bring the whole place back to life. You can see from Figure 15 that the effective water area was increased considerably by clearing the margins.

Two problems remained:

(a) A central very shallow area choked with reedmace – to some extent rafted (Figure 15) rather than rooted. Removal of those required ropes and a tractor. Manpower itself was quite ineffectual and we estimated that it would take twenty men working for a week up to their waists in water to do what a tractor could do much more quickly.

(b) The average depth was only three to four feet with no water deeper than five feet. This problem was, and is, insoluble simply because a digger would not only cut out the gravel but might well cut through a thin clay seam above the bedrock which in this case is a highly porous sandstone. The water table in the sandstone is quite low (that in the gravel seam is a floating water table: Figure 1) and if we'd penetrated the sandstone in digging we could have pulled the plug on the whole lake. This simply goes to prove how sound is our advice in Chapter 4 about getting a geological survey done at an early stage.

Anglers will be increasingly fishing waters like Cygnets, so let these lessons be learned in theory: it's dreadful to learn them in practice.

55. A trickle of pig slurry from an adjacent farm into a small lake where the photographer stands. In 1983 the A.C.A. tackled a large number of pollutions from the same cause.

56. (Left) A good team can remove quite large sunken trees with ropes and pulley systems if a good anchor point is available.

57. (Right) Aqualung divers discovered this and several other large sunken trees. They are finally removed by tractor.

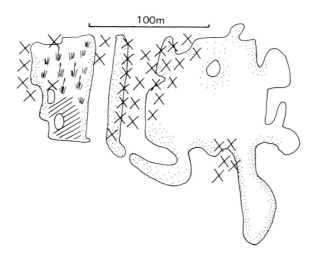

Fig. 15   Cygnets pond, showing the considerable surface area of the water gained by raking the foetid margins (stipple), clearing choking reedmace (diagonal shading) and too rich a growth of willow (crosses).

PIG LANE

This water benefited also from a geological survey. Again extra depth was preferable though in this case not absolutely necessary. The survey showed that below the bed of the lake – in gravel – was at least thirty feet of clay bedrock before a water-bearing sandstone layer. In fact the latter also had a good head of water so there was never any worry about losing the lake itself. However, we did have other problems to contend with. The easiest, because of past experience at Cygnets, was the heavily overgrown parts of the lake. Careful swim preparation coupled with swim raking and proper husbandry of the willow trees resulted, *within three months*, in a very attractive water.

A much thornier problem is the dragging of another lake, one edge of which will be within fifty yards of Pig Lane. It is not commonly realised that water flows through gravel as quickly as through a sponge. Pig Lane is on a gravel seam several miles long and if anyone digs a pit for gravel extraction anywhere on the seam, and pumps out the water into a drainage channel, then any established pit nearby will lose water. Figure 16 summarises the position. What *must* be ensured is that the water pumped out of the new pit goes quickly into the old one. Figure 16 depicts the procedure. Of course, the pumping never ceases because as fast as it is pumped out it flows in again through the gravel. But at least the old pit doesn't dry up, unless it is just too far away to pump to effectively. In any circumstance there is a critical distance beyond which you cannot pump: but beyond that distance the likelihood of being dried out decreases gradually. It's all very finely balanced and needs great care.

Fig. 16   The effect of pumping in the new, working gravel pit (right). Reducing the water level (the water table in this case) towards level 2, even briefly, results in water flowage through the gravel along the direction of the arrows, thus lowering the water level in the old pit (left) as well as locally lowering the water table.

58. This gravel pit is now flooded but the ridge-and-hollow working system, shown here before flooding, still exists in the lake. The tops of the ridges are shallow enough in places to take a good weed growth.

At Pig Lane there are two ways of going about it and a decision on this has not yet been reached. One is to pump the working gravel pit water directly into the old lake. The chemistry will be exactly the same, so no problem there. But the new water will be cloudy with suspended sediment. If the old water becomes choked with weed in the summer of 1983 then that is what we shall do, because nothing kills weed beds faster than a sandy/silt laden water mass. Against this must be set the risk of actually silting up the old lake. A second possibility is to pump on to the banks adjacent to the old lake so that water runs quickly into it yet is filtered by the intervening gravel beds. Anyone who doubts that water flows quickly in the manner described should fit a good depth record gauge on the lake. They will notice that within half-an-hour of pumping stopping the water level in the established lake begins to fall as the new lake fills up: as pumping starts up again the level in the established lake rises and that in the new lake falls (Figure 16).

Removal of excessive willows at both Cygnets and Pig Lane has resulted in an increase in marginal emergent plants which, by and large, are preferable to overhanging trees. The latter have, in fact, given us one of our biggest headaches at Pig Lane in the form of deeply sunk, waterlogged, willows. Some of these have been blown into the water during gales, others have been cut down and left in the water. Providing one can get a good rope around them, at depth, it is surprising how little manpower is needed to move them. But a diver is needed to set the ropes. What we have done is to tie up each underwater obstacle, buoy the rope at the surface, and then at leisure attach the buoyed rope to a bank rope and then to use either man plus pulleys or tractor power. Once again, we do not waste the logs obtained.

Although similar in some respects, Cygnets will never be a top-class water because it lacks the depth necessary in a small lake; but Pig Lane will eventually become a superlative fishery. Other examples of fisheries in which we have been involved with management will be mentioned more briefly at appropriate points elsewhere in the book. But the detailed examples given in this chapter surely serve to remind would-be managers just how wide awake they have to remain.

PART 3

FISHING ON SMALL WATERS

CHAPTER 8

LOWLAND STREAMS AND TRIBUTARIES

The problems with lowland streams stem from the facts that they often flow through good farmland (positions A and F in Figure 1) and that angling clubs usually realise that, though small, the fishing can be good. The R. Roding in Essex quickly springs to mind as a water appreciated by clubs; and the upper Great Ouse as one ignored by clubs, yet made famous by Richard Walker and his friends some years ago. In the first case you'd be pressed to secure some fishing, and hence to work on it, whereas in the second R.W. did secure a lease under fair terms and reasonable facility to work on the water. Ideally, you should look for the kind of water that has been neglected by clubs or where the farmer has had enough of clubs yet may be open to persuasion on behalf of a *few* anglers.

Many of these waters flow through what used to be water meadows, damp and rich in plant life. Now they may be cattle meadows which come right up to the water and include cattle drinks and well cropped banks. Work on the waters can, in fact, be quite minimal; the lopping of a bough here, the removal of a log jam there. But unlike our small ponds and mountain streams such work should always be done with the greatest caution: the lowland farmer is inclined to regard 'the odd bough' a lot more seriously than rampant willows around an old borrow pit. In the water itself it is often possible to do more. Weed beds can be trimmed, silt removed, the occasional diversion or dam introduced, or a raft constructed. On all lowland streams a *surface* raft of rubbish, in other words one that does not interfere with flow, is often the home of good chub and pike. Such rafts are quite easy to make, as outlined elsewhere.

Bankside cover can be a problem. Fine if the cattle do not graze to the bank top, perhaps hindered by barbed wire! In these circumstances purple willow herb will grow easily and can be used as cover – it seems to be the most common bankside plant on such waters apart from ragwort and dock (the last having dreadful seed heads for tangling line). It's quite easy to plant these and they do nobody any harm. In

112

59. An uprooted willow tree provides cover on a small pool on a tiny Yorkshire trout stream.

less constrained leases it is possible to plant willow cuttings which grow at great speed, but which are no real problem as far as trimming is concerned provided you are with them from the start and do not inherit someone's dereliction problem.

On Walker's upper Ouse we believe that all they did was the odd bit of trimming here and there, just enough to get a rod on the water whilst preserving cover for the angler. They also had true bulrush in the water in profusion – large beds through which and between which the constricted water often flowed strongly. In other parts of the country such a continuation of good water flow and bulrush do not occur. For example, many Yorkshire streams of the lowland, and those in the southern part of the Lake District, have few bulrushes. Bulrush beds are a great asset on any water provided they do not choke it completely. Grubbing them out is not easy but *is* effective in that they colonize only slowly, unlike the common reed by contrast. And they

60. Fishermen are conservationists. This tree trunk has been securely fastened in order that small mammals etc. may pass from one side of the stream to the other. It will also double as a debris trap for heavy material washed down during the winter.

prefer a gravel bottom and are thus both indications of bottom conditions and stabilizers of bed load. They also harbour a variety of food items which chub, in particular, and roach and perch really appreciate. As an aside, one might mention that Irish tench are partial to them too, though not, seemingly, the pike. They prefer the common reed beds.

One of the problems associated with chub fishing such waters is the kinship the chub clearly has with the damned plant! Not only does the fish often spend its time *in* it, or under an overhanging canopy of it, but on being hooked in open water will head for the nearest bulrush clump at high speed. We do not mean that it will head in the general direction of the bulrush, as would most species, but it will have as its objective a spot as deep into the rush bed as it can stick its head.

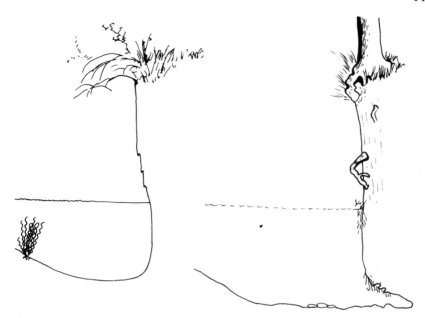

Fig. 17   (Left) Typical undercutting in silty banks of a stream.

Fig. 18   (Right) Undercutting typical of tough clay banks.

The bulrush is a great asset on any lowland stream or tributary. (In fact the two four-pound chub which Barrie caught on a small pond also had, unusually, a small clump of true bulrush for which they hightailed as fast as they could go). Planting is not easy but the stems and roots can be set in a nylon sack, with bricks and gravel, and the whole lot forced into a likely home in the gravel bed. As this has to be done in early July there's less chance of flood washing away the work before it gets established.

Another plus factor in lowland streams is that they flow quite often through valley alluvium, a kind of silt deposited over thousands of years by the migrating river courses. This forms undercut banks as the river course meanders, which it does, of course, much more frequently than the straight, hell-bent mountain stream in its rock strewn

A Fishery of Your Own

61. Re-stocking. Before release into a small lowland trout stream the fish are held for a few minutes in water from their new home. This prevents a sudden change of temperature after leaving the aerated tank in which they were transported.

progress. And undercut banks mean float fishing, laying-on, stret-pegging and trotting and all the other delightful things which go to make up small stream fishing.

Tackle control is much easier than on the turbulent hill stream, yet the undercut has to be understood. Sometimes it is as shown in Figure 17 if the banks are of alluvium. But if the bank is of tough clay the section in Figure 18 is more realistic. You wouldn't trot the stream in Figure 18: you'd touch ledge with strong line and hooks! If the stream runs through gravel beds you can expect *both* types of undercut to occur, so beware. And crayfish encourage the kind of undercut shown in Figure 18 by their habit of digging tunnels. This habit enables you to catch them for chub bait without waiting for the dark hours and a drop net full of fish pieces – simply stick your hand up one of the holes and pull them out. They pinch no harder than small crabs.

62. Although a correct and official tag, on a pike on a tiny Fenland drain, such tags are relatively inefficient. They are too small for pike much bigger than shown here and have a tendency to fall out.

The further one travels down small lowland streams or the tributaries of larger rivers the more embroiled one becomes with large angling clubs, and the more one's roving, stalking tactics become a matter of big river tactics and a long stay in one swim. The advantages of staying upstream are really that one acquires peaceful, yet varied fishing, for fewer fish but for bigger fish on the whole. Certainly the average size of fish is greater, particularly where chub, roach, dace and trout are concerned. Big bream occur less commonly, but they do occur: and when they find a suitable hole they stick to it. The trout are a result, often, of drifting down from some exclusive trout fishery, although in our scenario you can bait fish for them with equanimity. Included in a sprinkling of jack pike you may find the occasional very big one, or two at most: hooking them is like driving a tank down the village high street.

So you have great variety in the lowland streams, great fish, and good fishing if you can secure it and keep it quiet.

CHAPTER 9

MOUNTAIN STREAMS AND TRIBUTARIES

To possess a game fishing water is the stuff that pipe dreams are made of. In today's inflated world, fishing for trout, sea trout and salmon can cost a lot of hard cash even, as we said in our introduction, for that comparatively tame sport of casting a fly on an artificial put-and-take lake. But dream and reality need not be as widely separated as one may imagine, for there are a thousand streams and minor tributaries in the mountainous areas of this island which are largely unfished, and where an enterprising angler who is prepared to travel and work for his fish can organise a first-rate game water.

These fast flowing streams, often only a few feet across, present a challenge in every sense of the word. To find them one must leave a vehicle and take to Shanks' Pony, often climbing several hundred feet to reach, let alone cast from a particular bank. The rewards will not be enormous, at least in terms of weight per fish, and during the season you might expect trout averaging five to the pound if you are lucky; but rest assured that every fish, regardless of size, will fight to the end and taste like trout really should. Later in the year, where small steep streams reach more gentle valley slopes and merge together into minor tributaries there may come the opportunity to net sea trout, and even salmon.

Small mountain waters do not require a great deal of management or maintenance. During the winter months ample rainfall will swell them to a raging torrent that can erode, then deposit shale, gravel and rock over the bed and along the course of a stream until it may be altered out of all recognition. There is little one can do – or should want to do – whilst this upheaval takes place. Ken has often tried, but Mother Nature has always got the whip hand and destroyed his efforts by the end of the winter.

Summer months are vastly different. Although flash flooding may occasionally occur the usual problem facing the angler is that of too little water, hand-in-hand with an ever present lack of food to encourage and support fish life. With water possessing an usually acidic pH value (ie. < 7) it will quickly be seen that weed growth, prolific weed growth that can support and influence food supply, is a non-starter and what there is will generally be confined to members of the moss family, which somehow clings precariously to the bigger stones and boulders. That is if high nitrates, introduced under the guise of improving hill farming, has not reduced the moss to algal slime – but more of that later.

In fact a substantial amount of fish food will arrive in the form of insects and grubs that are dislodged from plants, bushes, trees, etc. that line the banks, with an addition in insect life that may be washed out from drains and dislodged at sheep or cattle crossings and drinks. So the obvious conclusions must be that management and improvement, where it is needed, should be directed towards this external source of food, and careful husbandry of all bankside vegetation, even down to humble clumps of nettles, thistles, various wild parsnips and meadowsweet to name but a few of the more common weeds, will be important.

Wholesale cutting out of overhanging branches and bushes to help one cast into an attractive pool or hole and an awkward fish that it contains will definitely be out, and even thinning to let light to a stretch where weed may be showing some signs of increase may often be a waste of time. Our experience has been that weed on this type of fishery will please itself where, and in what quantities it will grow, despite bland statements to the contrary made in some textbooks. One year there may be an area offering a lush covering of moss-like weed, on another that self-same stretch will be barren and attempts to replant of no avail even though one has made careful checks to ensure that no pollution or violent shift in the pH value of the water has taken place.

Branches and brushwood worth cutting out will be those that hang down *into* the water, forming a trap for debris carried along on floodwater. These natural surface barriers quickly divert water and often wash the tree or bush providing the obstruction out of the bank in a short space of time.

Undoubtedly the most useful job both from the point of providing a good head of fish and increasing their supply of food will be to construct stone wall dams. Dams, note, in the plural. Mountain streams have a fast fall as the ground drops away towards the valley and a single structure will only raise the water upstream behind it by a few inches, and then only for a short distance.

Four or five dams, perhaps even more, grouped along a straight between falls and pools already in existence can work wonders. Water that is held back will flood over dry land and wash out insects galore;

63. Ken's roving angling jacket. His objective is to pack into the pockets as much as he needs for the day's chubbing or trouting.

Ken did this on streams close to his home a dozen or more times during the season and never ceased to be amazed at the quantities of spider, grasshopper, bee, bugs and general creepy-crawlies that can be seen to drown and wash into the stream, eventually to finish in the pools below.

Naturally, the deeper water that is provided, especially where a hole is scoured below a dam, will attract fish and help to hold them and this is most important, for it is not generally realised just how far trout will travel both up and downstream in search of cover and safety. Small improvements such as these can be undertaken at the commencement of a season and will need but little attention during the following months. But where waters unite to form small tributaries in the valleys downstream a different picture will present itself, and management must differ accordingly.

Once water leaves the steep slopes its rate of flow will slow down, and silt that has been carried along will start to drop and then settle. Not only that, but the ground through which the tributary runs may also alter, with stone and rock replaced by areas of earth, clay and fine gravel. Again we are back into the world of erosion and deposition as the tributary cuts its way ahead, with silting a constant hazard that will need year-round attention by means of dams, siltboards etc.

Although the pH value of the water itself may not appear to have changed significantly since leaving the mountain slopes there may well be a significant weed growth through quite long lengths on some tributaries. This growth, totally unknown in many cases until very recent years, is the result of modern farm management which requires that large amounts of artificial nitrates be spread as a fertilizer over pastureland, to encourage grass growth. The trouble is that the resultant water weed may eventually become slimy algae that obliterates everything.

Washed along land drains and leached through the soil, nitrates will soon produce a 'crop' of weed for the angler. A mixed blessing perhaps, but one that should be exploited and accepted. Ken knows stretches where there had never been sight nor wave from a strand of greenstuff but which now supports heavy beds of water moss, myriophyllum, and water cress. He also knows of good tributaries where the self-same absorbed nitrates have reached such a level that

64. Natural and artificial crane fly compared. Heaven knows why the fish are fooled!

pollution point has been reached, the stream bed has slimed over and fish either killed, or have evacuated to other areas. Once this state is reached – before if possible – contact should be made with the Pollution Officer of the local Water Authority. We are aware that some Water Authorities are, at times, reluctant to reveal the nitrate figures in their region, and the sampling intervals are woefully infrequent.

As we said, it is worth trying to extend this weed growth and encourage it by the methods we outlined in Chapter 5. But control and management is often a gamble and should be accepted as such, for where one farmer practices intensive farming and the water bed is enriched by an even nitrogen supply, others believe in the 'leave-alone' principle or the 'slap a bit extra on for luck' theory and nothing grows accordingly. The pH, incidentally, can be changed from acid to alkaline by salting your water upstream with blocks of limestone or chalk, preferably the latter, or even something like Gault Clay if you can get it. Commercial lime works well, indeed better in the short term, but on any water you need regular supplies of a carefully calculated amount. We mention this in passing – please do not think we are advocating the haulage of tons of chalk etc. up mountainsides. But on the lower ends of some streams, where land levels out a little, it is quite feasible.

The fish to be caught in these small tributaries? Well, much will depend on what lies downstream of you. If there is a clear access from the main river then there will be a run of sea trout and salmon at the appropriate time of the season. But if an obstruction, man-made or natural, prevents fish from running up the tributary then the most one can hope for is a better class of trout than those which may be found in the headwaters above or, where there is a native stock, some excellent grayling. Catching them? Ah! a different and often difficult story.

Forget traditional fishing with a fly for instance. There is no room for chalk stream deliberation and fancy rod-work on these small waters; at best you may find an odd stretch where a conventional cast is possible, but more often than not lack of space to back or even forecast will rule out switch, spey, side and the steeple delivery. But there are more ways than one of creeling a fish and some of the alternatives open include dear old Isaac's method of dapping both the artificial and the natural. It is also possible to float a fly downstream over quite a long distance and to work one, sink and draw, with the aid of lead and a fine line.

The worm, that humble garden wriggler which has started so many of us on an angling career, is probably the best all-rounder on these fast waters, with an ample range of leads to help control and present them.

65. The fruit of hard labour. Trout from a mountain stream. Note the small selection of flies in the fly box!

66. Typical mountain troutstream water. Nothing here in the way of improvement is needed other than some selective branch tying to allow casting.

Finding and carrying worms, especially during summer months, is a problem, and one that the holiday angler in particular must constantly keep in mind. Slugs are easier to find and almost as good under spate, but less effective than worm under low water conditions.

Make no mistake, worming, the art of trotting the lobworm or a brandling through water so that it looks like a drain-drowned natural, is hard. But it is harder still to work that worm when the stream boils in flood. Weight, knowing just how much to use and where it should be placed – perhaps as a paternoster or a rolling leger – is the secret to successful worming. More especially there is a need to develop a sixth-sense that knows where the bait is at any given time, and how it is behaving.

Few people attempt spinning on these difficult waters, the probable reason being that one can lose a fair amount of tackle in the attempt – no slight matter with tackle at today's prices. But where the water is wide enough to allow an artificial to show some action on a retrieve, even if it be only for a matter of a few feet, then the angler must be prepared to spin if he wants the better fish – especially those back-end migrants. Cost can be reduced by making spinners, but the best 'spinner' of all is the natural – especially the minnow – which is available for the effort of catching.

The deadly drop-minnow, or even the spun minnow worked à la Alexander Wanless with fine trebles and a little weight, will both be winners on every fishery where they can be cast, with the accuracy of a rifle, and worked to perfection in odd corners that could never receive fly or a worm.

CHAPTER 10

PONDS

We grew up with these, so it is fitting that there is now a drift back toward them. Many ponds, of say up to forty yards diameter, are stuffed with stunted fish of one kind or another. But they need not be, and in management terms there are effective things you can do fairly quickly before getting too involved with fishing. In the first place make sure the water is not too overhung with trees – you can manage without marginal trees at all! Remove marginal trees and plant others well back from the water, but not all round it. Secondly, make sure there is some deep water, by dredging if necessary. Half a day with a J.C.B. will repay the money spent by the quality of future angling: leave some weedy shallows, obviously. Thirdly, if the water does have a huge head of stunted roach, crucians, rudd or whatever, organise the removal of a large quantity – someone with a pike water will take them off your hands provided the Water Authority will approve the move of *stunted* fish anywhere at all.

Anglers show a remarkable lack of initiative when it comes to stocking ponds. Barrie used to fish one of fifty yards maximum diameter that had chub in it up to four pounds – and he caught two of them. Another pond in East Yorkshire had a diameter of only twenty yards, yet had a dozen chub in it averaging 1-3 pounds. Chub do not usually breed in stillwater, certainly not in small ponds, yet they provide good sport, keep down stocks of stunted roach and themselves grow quite big. Similarly, you can try dace instead of roach for much the same reasons, except that they are not predators. They grow big: Barrie had a 14 oz fish a couple of seasons ago from a small carp lake into which he introduced chub and dace. The odd trout, perch and pike do not go amiss in a small pond, whilst it is fairly obvious that carp will grow big if the food supply is there. In the College pond mentioned in connection with polythene weed cleaning, there are numerous carp averaging 6 lbs, exceeding 10 lbs, yet the depth is a maximum of 5 feet and the dimensions 10 yards by 70 yards. Perch exceed 1 lb in the same water, rudd $1\frac{1}{2}$ lbs, roach 12 oz and eels 3 lbs.

67. Despite the weed this small Sussex pond holds a good head of crucian carp. They are under attack from a roach-pole – an ideal weapon in many respects. There are thousands of these small waters across southern England, each carved out of clay that was baked on the spot to make bricks for the construction of farms and dwellings nearby the water.

68. Modern technology. This Sussex dew pond set on the Downs has been cemented in. Traditionally, such ponds are clay puddled, but droughts in the past few years (note how low the water is here, in summer 1983) have caused the beds of many to crack and leak. Hence this modern replacement.

By contrast the worst kind of pond is completely surrounded by dense willows, with branches dead and dropping, with margins choked with blackened leaves and silt and with a central patch of thick hornwort, most of the time in a half rotten condition. During drought conditions the fish are seen gasping on the top, and it must be quickly apparent that the face-lift for this very common type of farmer's pond is to aerate it – let the light and wind in, and deepen it.

The only real problems with fishing these tiny ponds is the presence of other anglers. Ideally, there should be none! Try to fish them when no-one else is present and then employ the kind of watercraft we outline in Chapter 15. Tackle itself need not be all that important: the finest match tackle for the smaller species to the best of specimen hunting gear for chub, carp or pike. Always keep well back from the water's edge – remember that fish in a small pond can visit the edge by the merest flip of their tail! Employing a keepnet in a pond is a fairly conspicuous exercise – the keepnet seems enormous in the water. What

69. The reward for sensible weed planting and maintenance. Floating crust has accounted for a fair-sized carp sheltering close to the bank under the lilies.

70. (Left) An East Yorkshire pond with encroaching greater reedmace and collapsed willows.

71. (Right) A pristine chub of 4 lbs 2 oz. caught by Barrie in a tiny East Yorkshire clay pit on the tackle illustrated on page 155. The spoon is red coloured on the inside.

you really need is another pond nearby, an experience we have had on a surprising number of occasions. It is preferable to use a keepnet simply because returning a sizeable fish to a very small water can have repercussions as it passes on its disturbance to its fellow incumbents. In bigger waters it may not matter so much.

Another serious question for the pond enthusiasts is how to groundbait and what to use. If you happen to be fairly casual you can go a long way without baiting up at all. After all, the fish should find you before long. But even in small ponds it is soon obvious that fish have preferred spots, and definite patrol routes. So to ambush a chub or carp with free offering of the hook bait is good tactics, as is feeding maggots or sweetcorn into a small pocket in the hope of getting the roach or crucians going. In much greater doubt is the wisdom of using much cereal groundbait. One can imagine just how quickly a shoal could get choked off! Cloudbait is a different question and, coupled with maggots, can be most effective. We describe a somewhat specialised use of cloudbait in Chapter 16.

We have fished and worked ponds so small that we have almost begun to feel foolish. Almost, but not quite. And the enthusiasm of some local passerby when you have in the net what they regard as a monster is quite funny. Or, at least, it's funny at the time, but not when you find half the village fishing the water the next time you go.

CHAPTER 11

MARSHLAND DITCHES AND DRAINS

Ken was brought up on ditches so small you could scarce wash a cat in one. In fact the Pevensey Marshes where Ken spent his childhood are probably typical of many marshland areas – either in whole or part – that are spread through the country, so it may be as well if we enlarge slightly on what they do and how they do it.

The marshes, prime grazing land for store cattle, have to be drained and kept so if they are to be of any use to the farmer. In rough outline the system demands that land-drains a few inches in circumference be laid under and across each enormous meadow to bring water into ditches set at every field's edge, each of which is up to six feet or so in width. These then lead to and merge with larger drains (Guts is just one local name for them) and so on until eventually large sluices, or Havens, are established, each of which resembles a small canal in both shape and size.

It is these wide straight Havens that drain the water towards, and eventually into the sea, either by means of pumps or by tide drains that open whilst the tide is low, closing by means of a non-return valve when the tide flows in again. Seen from the air, the whole area has the appearance of a large herring skeleton with the backbone represented by sluices, whilst the ribs become ditches that run from them.

In theory the area should be a fishing paradise. In practice the waters always support a fair head of coarse fish but seldom achieve their potential; failure stems principally from our old enemy water flow, the amount and speed of water passing along the various drainage channels. As we said, drainage starts across the huge marshland meadows by means of small earthen land-drains set in such a manner that there is a fall, or slope, leading the water into the ditches. All well and good, but from there on trouble starts.

Because the marshes are dead flat there is no further fall and subsequent movement from the water is nearly non-existent. So silt, carried in suspension from the fields, quickly falls to the bed of the ditches and drains, filling and blocking them in a relatively short space of time. That, of course, is why one so frequently sees a mechanical digger at work on the marshes, scooping and depositing mountains of black earth beside the banks.

Enter now a further problem. The flat marshes are devoid of trees and, in most cases, even bushes of any significance, which simply means that little or no shade is provided during warm months of the year. Already the recipe for an oxygen hazard is under way; no water movement, silting, no shade to prevent the water from warming up, so both bankside and submerged water weed will grow and multiply until the smaller waterways become choked and bigger waters seriously overcrowded.

Of course, the picture changes in winter. Then weed dies back, rain will bring excess water into the ditches and drains that will be cleared by pumps and sluices, tending to speed up current along all water courses. But the flow is always pulled, never pushed, so its strength will be insufficient to cut away silt which leaves the original problem, a silt build-up that can only be tackled by dredging, repeated so that each length of water can expect to be shorn bare every third year.

Maintenance and improvements that can be made by the angler under these circumstances? Well, physical improvement by dam construction is time wasted. There will be little enough flow at the best of time to make it work and they will be sure to upset those farmers whose whole time is spent on draining water, not holding it back.

The angler can only concentrate on a form of maintenance by clearing the weed where it is at its thickest, especially where waters meet and merge with larger drains. By doing this some small oxygen balance can be restored and fish encouraged to move into smaller waters where concentrated efforts for their capture can be made. Carp and tench, in particular, are easier taken in a small clearing that can be accurately groundbaited, and winter pike deadbaiting along marsh ditches and drains can provide excellent sport.

Beside being prolific, weed growth on the marshes can often embrace many varieties and on a recent count Ken discovered marsh marigold, willow herb, meadowsweet, purple loostrife, reed mace and bulrushes at the margins, with broadleaved pondweed, Canadian pondweed, water milfoils, hornwort and, of course, duck weed aplenty below and on the surface.

The best means of hitting the growth is by raking, either with the long handled hay-rake or Barrie's mobile swim-clearer. Of course, marginal growth can be hand-pulled or lifted with help from a fork, but on no account attempt to enter the water bodily; narrow though the banks may be and small though the water may seem, the bottom layer

72. After the dredger. This scene on the Pevensey Marshes is typical and shows the traditional steep bank, with a cattle drink cut out in the foreground. Weed and bank growth has been removed and left along the side, but enough weed of the bulrush and reedmace type will soon grow again from roots which stretch far into the sides of the bank.

of silt can often be head-high and pull like a demented dentist once you are stuck.

As would be expected, aquatic life is on a par with the density of the weed, and insect larvae of every description; worms, especially leeches, crustaceans and molluscs simply drop or crawl from every pile of weed as it is freed and landed. In fact, Ken has more than once resorted to those marshland ditches to collect a supply of snails and caddis larvae, together with weed, so that he could stock or boost another fishery.

What applies to the Pevensey Marshes also applies to the Fenland drainage system to varying degrees. The farmers want as much water as possible held back in summer (to reach the crop roots). It is not our place here to expand on how crazy is the land drainage of our nation: for the moment the angler has to put up with it. And in the fens the

73. This long, straight, tiny, shallow Fenland drain, full of good fish, is part of a 'closed' system which can be managed successfully.

angler is not really a manager unless he belongs to a club which leases or owns a stretch of drain or river (the two terms being almost synonymous these days). The amount of work one is allowed to do is relatively minimal, but weed clearance is certainly one. Floating weed often accumulates at junctions of drains where eddies abound and, in the areas of the eddies incidentally, heavy silting may occur. Floating weed can be removed with the shallow net described in Chapter 5 for weed trapping. Or, at least, it can be held in abeyance. We used to do that whilst fishing for chub on Barmston Drain, near Hull. Otherwise the standard weedcutting can be done. Strangely, in Fenland (as opposed to the Pevensey Marshes) quite a lot of the drains have a firm bottom and it *is* possible to wade and cut. Remember, we are talking all the time about *small* waters in this book, not ten foot deep drains! When wading and cutting beware of cutting your feet and legs! Even in summer your legs may become chilled and a cut, even a deep one, can go unnoticed.

The actual fishing of the tiny drains and ditches can be mind bending. We know of no Fenland drain over eight feet wide that does not contain twenty-pound pike. When a water of this size will grow such pike you can rest assured that big cyprinids will also occur. Strangely enough, each of the very tiny drains we know usually has one cyprinid which does better than the rest in the water and, perhaps, the roach is most commonly that species. However, we know of tiny drains both in Fenland and elsewhere that hold big carp, chub, bream and tench, but usually only one of these is dominant as a big grower. If the carp are big, the tench usually miss out, and so on. Several such drains are good rudd waters and we used to fish one (today unhappily in the foundations of a now derelict greyhound track) that yield rudd over 3 lbs and numbers of 2 lbs. Several of Barrie's largest rudd came from this water, to floating or slow sinking crust. On another water we took big chub from down thirty yard long culverts, little more than five feet wide and eighteen inches deep. It's rather like a quieter version of mountain stream angling (Chapter 9) in that you must really explore, observe and poke your rod and bait here and there. These drains go in cycles, too. One which used to produce big bream is, in 1982, producing big tench (over 6 lbs) in water less than two feet deep in a drain, perhaps, twenty-five feet wide.

Approach work is a real problem because such ditch-like waters commonly have steep banks close to the water's edge. In summer there will be an abundance, even super-abundance, of weed and, perhaps, some marginal emergents to give you cover; and in winter the water can be coloured and deeper, but always it is difficult getting near the very big fish. This is so true that the big fish potential of many waters of this stamp is totally overlooked. What is more, some Water Authorities and local drainage boards show no imagination whatsoever as far as stocking is concerned. Waters that could be producing big carp, tench, chub and other species, as well as supplying small cyprinids as stock for their depleted big waters, are left to the occasional renegade stocking by anglers or the accidental stocking by nature.

Tackle on such waters is far more varied than on the mountain streams: indeed, the only real comparison between the two is the width of the water! Tackle has to be varied if, on the one hand, you are to expect pike over 20 pounds, and on the other a netful of roach over the 1 pound mark. In essence, as the tackle is no different to fishing any other (larger) lowland water we need not enlarge on it here, but merely emphasise that the approach work has to be first-class. You cannot judge such a water, for example, by a close season walk along the top of

74. Baiting up for tench on the Old Bedford River. Even on small waters heavy baiting up may be necessary to hold the fish.

the bank! You'll see nothing. Rather, the exploration is done by fishing. Try to work out where the big fish might be – as we did with those chub – get in position early, keep quiet and fish. If they don't come to you after you've fished, try somewhere else. Rudd, in particular, are often in the most unlikely places: they go so far up some ditches they are almost in the crops!

Quite recently Barrie discovered, along with his colleague Colin Goodge, a Fenland drain of the most striking proportions. At its widest it is, perhaps, twenty feet – may be twenty-four feet – and at this place is no less than sixteen and a half feet deep! Where the width reaches under eight feet the depth is almost the same. It's packed with big roach, including fish well over 2 lbs. Now here is something for the Water Authorities to ponder on, isn't it? Instead of having drains about two to six feet deep, and then running them virtually dry every winter, ruining the anglers' sport in the process, why not have them fifteen feet deep and have them to, say, ten feet in times of flood?

135

Admittedly, they would silt up in time, but they do that anyway. The frequency with which the Anglian Water Authority dredge the Old Bedford 'River' suggests it would be more efficient to build their H.Q. on the bankside. If they opted for such depths they would have less of a weed problem and they would not have to go to the expense of trying to sell us grass carp.

PART 4

FUNDAMENTALS OF SMALL WATER FISHING

CHAPTER 12

TACKLE

In our opinion small waters make more demands on fishing tackle than any that we know. If you don't believe us, then take a look at some of the pictures in this book which show every conceivable type of small water that you could imagine. They include tiny ponds and rivers around the Home Counties where stealth and concealment are the order of the day and delicate tackle reflects the only approach, through the extremes of wild mountain streams and isolated lakes in near-unhabited parts of the country, where a hard slog to reach the water will be followed by angling gymnastics before baiting begins.

In every case that we have illustrated (or described for that matter) the tackle to be used must be able to withstand hard wear, constant use, travelling hazards, extremes of weather and unintentional as well as occasional intentional abuse. All that, of course, must be added to the normal pre-requisites that apply to tackle used in general fishing, namely that it must enable an angler to reach, hook, play and land good fish by all legal means regardless of conditions either on the bank or in the water, and continue to do so.

Naturally, those considerations spread right across the board and are not just confined to hardware in the rod and reel department. To fish successfully one must be able to fish comfortably, no small order in this country of extremes in weather. Keeping warm and dry helps to promote concentration, which means that clothing, footwear, even bags in which to carry tackle should all receive some pretty demanding scrutiny and consideration.

You would be correct in thinking that only the best, the very best, will do when you fish small waters, but the best does not necessarily mean the most expensive. For years we responded to the spell of top names in the tackle world. Then we wondered whether it might not make economic sense to buy more cheaply and, perhaps, replace some items a little more frequently than would be normal and we started studying the catalogues of *all* the manufacturers and selecting from them what appeared to be sound buys.

Surprisingly, we have not only found a great deal of difference in the cash situation – no small consideration today – but we have made a further discovery that there is often little difference in the wear factor on many items when compared to the top named brands. Even further, we have also discovered a completely different approach on our part through fishing with these cheaper items. As an instance, we took to purchasing mill-end fly lines which were pounds less, still efficient, and lasted us several seasons. Yes, the expensive line we had previously used would have lasted longer, but because of its price we were conscious for its safety all the time we were fishing and would refrain from trying a 'dodgy' cast into a tight situation where it might possibly be damaged. Small water fishing is all about taking risks for fish and we soon realised that when you are hanging back because of your tackle then good sport, the very essence of small water work, is lost.

Keep your tackle strong, light and durable. Don't regard it as a form of male jewellery, something to be worn and admired, and you won't go far wrong. It is worth remembering the story Ken heard about the two anglers arguing in a tackle shop about the merits of two different landing nets. After ten minutes of wrangling, along came a third angler who coldly informed them that providing it would do the job properly, he would be quite happy to use an enamel chamber-pot with which to land his fish. What did they want? To land a fish or play charades?

TACKLE FOR GAME FISHING

It would be a brave man who sat down and prescribed a complete range of tackle for small water game fishing – at least, one which was contained within the covers of a single book. The three accepted styles of game fishing – fly, spinner and worming – are art forms that naturally require a variety of rods, reels and lines if they are to be practised with any degree of proficiency. Water conditions usually ensure that only one style of angling may be practised at any one time, i.e. spinning in heavy water, the worm as water clears and runs off, so the tendency is to form separate 'sets' of tackle which allows the angler to be ready for any single type of water contingency. But even when these 'sets' of tackle have been made ready, the angler will find that he often does not appear to have the right set to hand.

If he travels extensively for his fishing he will encounter enormous geographical and ecological differences in waters that are spread across

75. The quickest way to contour a new lake, whatever its size. Adjusted carefully, The Seafarer distinguishes also the nature of the bottom and the position of weedbeds.

the country, a formula that ensures no complete tackle cover able to embrace *every* type of water. The angler on the Yorkshire Dales stream with his long fly rod which enables him to keep back from the unencumbered banks becomes completely lost when faced with the confines of a small mountain or lowland stream, where the angler must wield a toothpick rod under bushes and low overhanging trees.

So we offer no directions or commands in this section, leaving a final choice to the common sense of the individual. We do, however, pass on some lessons from our experience, most of which were learned the hard way!

RODS

Long or short? Thin or thick? Fast or slow action? Single or double handed? A suitable weight for length ratio? These are just a few of the questions that must pass through the mind of an angler who is contemplating a new rod. In general, the water he will fish and more especially the flora that surrounds it will decide many of his problems and preclude some of the choice. We set down on paper many thoughts about rods, their action and how to choose them, when we wrote our book, *Plugs and Plug Fishing*. Admittedly, our thoughts were directed to bait casting rods at the time – but what we wrote applies to casting with both natural and artificial bait of any sort.

We hope that the advice we offered prevented a good many anglers from making the mistake of choosing a rod *for the rod's sake*; perhaps because it is made from this or that material, because it has this or that type of fitting or it bears that particular brand name – all items that can only detract from the job in hand, which is finding the rod which will get a lure to a fish, play, and then land it.

In the main we make our own rods, choosing inexpensive blanks, concentrating on first-rate fittings, and then using the rod, and using it hard. If it breaks we set to and make another, annoyed with the inconvenience but never with the actual breakage. As we said earlier, and will repeat at regular intervals – if you want the best out of small water fishing, you must take chances with your tackle.

No one rod will do all things. But there are times when one rod can be coaxed, by adjustment, to do several tasks. We are not talking of the Combination Rod so beloved of our Victorian ancestors, where a battery of top joints were harnessed to lower sections to stiffen or slacken action. Instead, we refer to a general purpose rod based on a traditional fly rod, a blank which can be made to 'double' for worming or light spinning by the addition of a butt section.

Ken hit on the idea of such a rod when he described the ultra-light spinning rod in our book *Spinners, Spoons and Wobbled Baits*. This basically was a Fibatube 4/5 fly blank with a couple of inches lopped off the top joint, and a plain handle with sliding fittings fitted to accommodate a light-weight fixed spool reel. It was a great success and led to Ken's latest effort which was made for him by Vic Gibson at Dons of Edmonton. It consists of the same weight of rod blank without anything cut from the top joint, but fitted with screw fly winch fittings. To the base of this can be added an extension handle (custom made) which gives an extra 15 inches on the rod's length and allows for a small fixed spool reel to be added. The extended butt puts the reel in the right place for casting, as Photo 76 shows.

With this, Ken has successfully hauled innumerable trout from small Welsh and Southern Counties streams and rivers, plus a good few sea trout, and a couple of heart-stopping grilse. It would also be fair to confess that he has been broken a few times.

The rod need not be in the size and weight that Ken has produced –

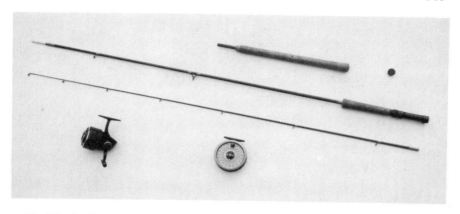

76. Ken's all-purpose rod. This 7 ft 4/5 fly rod has screw winch fitting and a push-in extension butt. As a result, he can fly, spin or worm fish simply by changing the reel. The rod has handled sewin up to 5 lbs, pike to 8 lbs, and stream trout galore.

indeed, a friend of Ken's has had an 8½ foot No. 6 made on the same principle and this fishes remarkably well on quite large open rivers. Like its small predecessor, when spinning or worming, the thin tip seems to reflect every move the bait makes underwater and, with care, quite heavy spinning lures can be made to work efficiently.

REELS

There is a wide selection of different fly reels on the market at the moment. Some are big and ugly – some are small and equally repulsive. Most, in our opinion, have a ratchet action that makes far too much noise, but more important than that is the fact that often there is too much play between spindle and spool. Choose one with as little play as possible and don't regret the extra few pence that such an item will cost you.

A reel small enough to slip into the pocket is a godsend when you want to change fishing styles or shift to a different weight of line (fortunately a fairly rare occurrence). We tend to opt for those models with an internal rim rather than with an external flange. Oh yes, it's great to put pressure on a running fish from the rim of the reel, but not

so funny to try and clear a high-spot when you have dropped the reel
on stones and dented the rim to the body. We have long and bitter
memories!

For spinning and worming it has to be a fixed spool reel. There are a
few enthusiasts who use a spare fly spool loaded with monofilament
and change to that when they change styles, casting by pulling and
releasing loops from between the rod rings, but by and large the age of
the centrepin is dead.

There can be few things more efficient or pleasant to use than the
modern fixed spool reel, especially where a few ball-bearings have been
incorporated into the design. Skirted reels are worth the extra expense
– try a day of casting through pouring rain and a howling gale with an
unskirted model and you will see exactly what we mean.

More recently Ken has been converted to the closed face reel, in
particular one of the lightweight match fishing models from which a
fish can be played off the handle. It is lazy, idle, morally indefensible
and grossly inefficient from the point of line strain – but a delight to use
where the stream or river is jammed with trees, bushes and high
vegetation. Ken has definitely not regretted breaking into his piggy-
bank to buy one.

LINES

If it is monofilament, buy it in bulk from a reputable dealer who
receives regular deliveries from the manufacturer and you won't go far
wrong. Most monofilament faults arise from exposure to excessive
heat, ultra-violet rays and overstretching, either when knotting, or
retrieving a snagged lure of one sort or another. It is still a cheap
commodity – work out the hours of use/price ratio for a spoolfull and
you will see what we mean. If you renew lines all round three times a
year you are on a profit margin both fish-wise and nerve-wise.

Fly lines we have already mentioned. It depends whether you are a
user or a collector in our view – but we are not dogmatic about cheap
mill-ends. We are, however, sure that the average angler using small
waters gives himself endless effort, heart strain and nervous debility by
sticking to a double tapered line. It's a subject to which we shall return
later, but consider carefully the water on which you will regularly fish.
Chances are it will be narrow and will require little in the way of
distance casting, which means that the normal double taper line just
does not become efficient at any time – the weight in the belly of the
line constantly remains on the reel. Substitute a weight-forward in the
self-same conditions and you start with enough weight to make the rod
work, bringing vastly different results.

Floating? Slow and fast sink? Sink tip? Well, again it depends on the water. We find most of our work is done with a floater, and by a little skilful use of cast length find a genuine need for a sinking line on only a handful of outings.

There are many things which we carry and never seem to question or use. Some are 'musts', some 'might come in useful', others we have carried so often we would miss them in the bag, and worry about their absence. We have prepared the following section with our hands on our hearts and we list the bare essentials we carry. To prove our point, Ken shows his outfit for fly, spinning and worming tactics on overgrown streams in Photos 77 and 78.

LANDING NETS

Necessary, but often evil. We do know anglers who never use one and rely on beaching fish; strange to relate they never admit to losing a fish, either. We are not that expert, if that is the right word. We both use them, Ken a triangular shaped one which he favours because he feels that the flat front slides easily under a fish, especially in shallow water, while Barrie uses a round one which he insists is more efficient and suitable for all occasions.

You can string them round your shoulders, hang them from your side, and generally encumber yourself with them in umpteen ways. But really, with a small water, one is hardly likely to spend long periods wading out in the stream and the majority of one's work will take place from the bank. A long-handled net with spiked end that can be carried and set down is not such an awkward tool as may be imagined and we find it fits the bill admirably. There are models on the market which allow you a set of screw-in extension handles so that you can cover every situation so to speak, besides rodding drains and cleaning chimneys with them during the closed season. We feel them to be gimmicky and definitely evil. A handle is a handle and its only other task should be that of being able to support one as a wading staff should it be necessary in an emergency.

WADERS

Fancy green ones look charming and tend to 'fade away' with exposure to the elements. Black ones (as those worn by sewerage workers etc.) are tougher, cheaper and last longer, especially around the knee area where as much wear will occur as on the sole end. Kneeling is a necessity and those who deplore waders wear breeks, heavy boots, and use a set of kneeling pads. Ken even came across one enthusiast the other season wearing brogues and plus fours, who was wading up to his trunnions (just below the hip pockets, to the uninitiated) in cold water.

Now for the bags, creels, special jackets and boxes in which to place or hang the items we consider to be essential. Easiest way for us to handle this section is by setting out the options and then feed-in our thoughts and experiences.

BAGS

Large ones are convenient and consume tackle, changes of socks, food, fish that may be caught and other odds and bobs with ease. They keep stuff dry, but are difficult to carry over long distances unless you constantly shift the weight from the strap off one shoulder to the other. They are also very expensive. Barrie converted Ken into using a hiker-type rucksack several years ago and when he is set on a long, hard day, it is this he loads in favour of anything else.

CREELS

Pretty, traditional, lightweight and definitely the 'in' thing along some fisheries on one hand; clumsy, restrictive (they don't 'give' like a bag) and not in the least waterproof on the other. On balance they are fine for an evening's fly fishing, but hardly the thing for a rough day's work.

JACKETS

There are a number of them on the market, and with a plethora of pockets, net rings, clasps and buckles that adorn them they are certainly capable of shifting a fair amount of tackle around the body. Ken uses one pretty exclusively even when he coarse fishes, but admits to being a glutton when it comes to the sandwich stakes, so takes a rucksack with grub and spare clothing on a long day, setting the bag down around the half-way mark on the fishery and working back to it.

BOXES

Not the type with a carrying strap, but the innumerable boxes we seem to accumulate for flies, spinners, worms, weights, casts, hooks, and so on. We covered many aspects of problems that centre around carrying small objects in our book *Tackle Making*. There we warned of the 'box within a box' syndrome, and the habit of buying too large a box, then feeling obliged to fill it with something!

Small plastic boxes are the lightest but not necessarily the cheapest, though with a little imagination in today's plastic packaged world you can usually adapt something into good use. Ken uses slim screw-fit plastic boxes that once housed camera lens filters to hold lead weights and hooks for worming.

The idea par excellence is to custom-build your own boxes to hold tackle, which is relatively inexpensive and adequately described by Barrie in a chapter of the afore-mentioned book.

MISCELLANEOUS ITEMS

Well, there are thermometers, scales, marrow spoons, tackle retrievers (otters), sunglasses, knives, scissors, ornamental key rings, left-handed can openers – to name just a few. The more you carry, the heavier it becomes, the less inclined you are to fish and concentrate, and the day becomes wasted.

We all have favourite items of equipment, some of which are talismen and bring luck, others we fondly imagine *might* come in useful. There is little that cannot be improvised in an emergency, however, and the man who travels light is inevitably the one who brings home the best bags of fish – or at least, so we have found. It's all a question of sorting out your own particular evils.

TACKLE FOR COARSE FISHING

There is no place on small waters for beachcasting equipment, or any pokey rods for pike fishing, for example. Since most of the waters will be less than two hundred yards across it would be unusual to need to cast half that distance. In many cases deadbaits can be swung out gently, or even cast by hand or with a casting-stick, thus allowing the use of even lighter weight pike rods such as the Mk IV carp rod. The same delightful rods, in cane or hollow glass, can be used in carp fishing (again there is no need for giant leads, stiff rods and distance casting). For perch, tench or chub fishing, Mk IVs are fine, but the lighter Avon version is probably a nice tool to use, as it would be for zander, medium sized carp and bream. For other species of coarse fish – bream, roach and rudd – there is a lot to be said for going over to match fishing tackle, not usually to short rods and sundry tips, or poles, which are largely unnecessary on small overgrown lakes if not on drains – but to lightweight glass and carbon rods, quite long, and fine line with small hooks. The most marvellous sport can be had with roach on small lakes particularly if you can locate a shoal in deepish water against a willow which trails out over the water. And since the

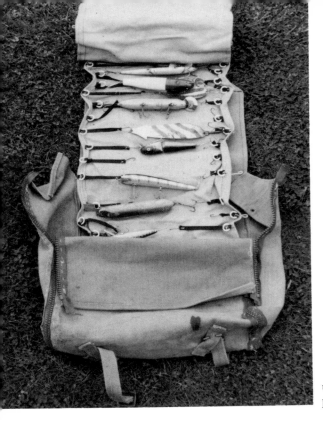

77. Prototype roll-up lure and tackle bag, shown in partly open position with lures accessible.

fish will be close to you, fine line, and hence the better match rods, are the things to use. Floats can be chosen according to the circumstances.

Of all aspects of tackle on small waters, notwithstanding the durability factor mentioned at the beginning of the chapter, the most important is drabness. Rods should *not* flash with thick shining varnish, nor should landing net handles. All, without exception, should be dull, painted matt green, brown or black, or a manufacturer's product dulled down firmly with a carborundum paste. The small water angler is so close to the fish that he must merge into the background. If you follow advice elsewhere in this tome that background will include emergent marginal weeds just in front of the angler, and shrubs and trees behind him and away from the water's edge. Again, ideally, the general ground level should be very little above that of the water.

Remaining items of gear are much the same as for any other fishing,

148

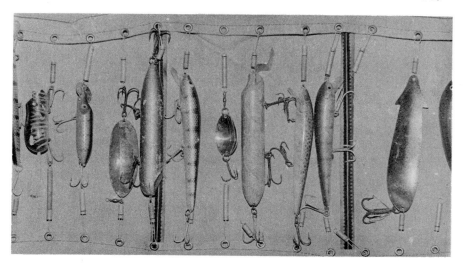

78. Detail of the roll-up lure bag with varied plugs and spoons in position.

though it is imperative that big keepnets are not only of soft material but very dark in colour and dull in texture. We have often laid-on *directly alongside* a twelve-foot keepnet and taken fish steadily!

As far as game fishing tackle goes, or wandering a tiny drain in search of pike, it is obvious that weight, or the lack of it, is vital. The game angler traditionally has carried a shoulder bag, landing net or gaff, and his rod (and hip flask). And really one should have nothing more in terms of *items*. However, what you have to carry in the shoulder bag for a full day in mountain country, or miles away in the fens, requires some thought. Barrie recently invented a haversack which although designed primarily with the lure-mad pike or salmon angler in mind, is very successful for any wandering angler. It works on the principle that in one's hand is a rod and a net: everything else goes on the back, in either a shoulder bag or back pack style, the latter being Barrie's preference.

The photographs are to some extent self-explanatory. It is assumed that the most common reason for opening the bag during the day is to take out and replace a lure. So the lure roll-up container is fastened near the flap. You open the flap, unroll the holder, and there you have, perfectly displayed, all the lures you happen to be carrying. Because of the unique clip and ring system not only are the lures untangled, but any length of lure, within reason, can be accommodated. Below the lure roll is a space for a thermos flask, food, waterproofs, and also unhooking equipment and camera. In fact, the only decision you have to worry about is the actual weight in relation to the distance you'll be travelling, cutting out items as appropriate. But for sheer convenience for the mobile angler there is nothing quite as good as this. Of course, you'll probably not want to use it on a small lake, but for fellside wanderings after brown trout, a day after salmon, or a day along a labyrinth of small Fenland drains, it is perfect.

Finally let us re-emphasise that for small water angling *all* tackle and gear, whether fishing, clothing, or brolly, should be dull, dark coloured and drab. At times it seems almost impossible to buy a dark brolly, but black and dark green ones can be obtained and you can colour them with Dylon dyes or with Mesowax, although the latter seems not now to come in the excellent dark brown version that was once available.

CHAPTER 13

COARSE FISHING

There is an historical perspective concerning coarse angling on small waters which we'd like to record simply because we began our angling days on these waters, fishing this style, as what are called today 'pleasure anglers'. When we were raw youngsters every village had its old angling sage or guru; a bevy of half-a-dozen or so, more or less regular, angling adults; and a similar-sized bunch of garrulous kids, in which we were included.

Angling techniques could hardly be called advanced. You always used floats (except in Yorkshire, where drilled bullets were often in use) and only two kinds of these: a crow quill or goose quill, and a perch bobber (cork body and straight, cane stem). The crow quill was for 'roach' fishing and the perch bobber for perch. And, of course, the Fishing Gazette pike bung, the main two criteria of which were that (a) it could be seen from the *next* village pond, two miles distant and (b) it fell off the line at the least opportunity.

At about this time, the 1950's for us, there sprang up the specimen hunting movement which goes from strength to strength today. All over the nation the garrulous kids faction began hunting the big ones. Even if things were not very productive at least they were quiet, because rule No. 1 of this new breed was to be 'a quiet and go a angling' (aided by rule No. 2 – try not to smile). Inevitably, the garrulous kids soon began to sneer at the sages – an activity directly proportional to the speed at which they could travel through the meadows carrying a full load of specimen hunter's tackle.

But there were some beneficial side effects. Because of the attempts to concentrate the mind and effort, and because the early specimen hunters were largely correct in their attitudes and approaches (even if such laudable attributes were somewhat lost in translation to the villagers) fish were caught, and big ones too. Barrie's roach catches of Chapter 16 were the best in that village for decades, whilst the consistent capture of big perch by the Hook Specimen Angling Club (no less) were good by any of today's standards. So a lot of good fishing, approach, and techniques actually came out of coarse fishing the small waters, whether it was chub on the upper Ouse or crucian carp on a flooded Kentish iron working.

And, concomitantly, the tackle evolved – the old three-piece whole cane and Lancewood rods being replaced quickly by Richard Walker Mk IVs and Avons in built cane. Free line fishing was the order of the day, and even in pike fishing free lining and light ledgering was developed with great success by Barrie (see *Fishing Big Pike,* Chaper 4). Today the specimen hunting attitude and approach still works well, in particular when you are chasing fish of several pounds upwards: tench, carp, chub, trout, perch, pike, and eels. The Mk IV built cane carp rods have been replaced by 10-11 foot glass or carbon fibre rods, and these need only be soft or through action rods. There really is no place for the specimen hunter's fast-taper rod in small water fishing, because rarely does one need to throw a concentrated weight a long way, or pick up a great length of line on the strike. The glass Mk IVs and their many approximate equivalents are beautiful rods to use and, except in very log-ridden swims, will probably handle any small water fish that swims.

Other tackle need be only what the specimen hunter normally uses: matched reels, a variety of breaking strains on numbers of spools; plenty of eyed and spade end hooks and the usual array of leads, forceps, weighing nets, slings, landing nets and so on. Even floats!

But there is another facet to this historical development. Whilst the specimen hunter was creeping around small waters the length and breadth of the U.K., others were learning new skills on the wide rivers and big drains – the match anglers. Match fishing reached new heights in the mid-late 1960's and may not yet have reached its zenith. As well as the misconceptions which always accompany a developing, expanding movement, there arose an appreciation of fine tackle, particularly the use of light lines (1-3 lbs b.s.) with superbly manufactured sets of floats, and hooks in the 24-18 range (tough wire, barbless, spades, attached and unattached). We also saw the development of short ledger rods; long, tip-action, glass rods; poles and finally carbon fibre wands of all kinds.

The effectiveness of such equipment on small water fishing for roach, rudd and bream has to be seen to be believed. I recently watched John Holmes of the Beecroft winter league match team take a 100 lb plus bag of roach and rudd, which included no less than twelve roach over 2 lbs. The tackle was a waggler carrying and locked by two swans up, and a dust down, and hooks changed at various times from 18-12 (the last in deference to the size of the roach). No specimen hunter could have expected to do better and none did, although a friend of mine had nine 2 lb. roach on the same water on the same day. However, that's *all* he caught: no 'make weight' fish at all.

So clearly the match angling approach works well. But what is really needed is a combination of the two. Be prepared to go in for 'snatching' tackle or carp bashing tackle if necessary. The principles of watercraft (Chapter 15) should be employed in both cases: by all means use a matchmen's plastic basket, but do not dunk it on the deck as those anglers did in Chapter 15. Put it down quietly; leave the emergent weed cover alone. Try to be flexible, using the tackle and techniques of both systems, as appropriate. But, above all on small waters, use watercraft.

Armed with the above-mentioned sophisticated equipment you can launch yourself with confidence into coarse fishing small waters – not to catch small fish, but to catch good fish and big bags.

There remain several matters upon which we'd like to expand, namely the use of floats, poles, modern terminal rigs and pike angling. Our thoughts on these do not upset the basic principles of coarse angling on small waters, which we have outlined above, but there have been relatively recent developments upon which we'd like to give guidance.

It follows from our comments on the match angling boom that there is a large selection of superbly manufactured floats available – wagglers, zoomers, sticks, Avons, etc., etc. Amongst the excellent design features are: that they are marked with the exact shot-carrying capacity; and that they can be obtained in dull, matt finishes with only the very tip of the float brightly painted. For small water fishing the actual *type* of float may matter much less than its shot-carrying capacity. Providing you can fish delicately, close in, and then change quickly to a float carrying several swan shot, in extreme cases, that is all that matters. Actually, stability of the float is rather less important unless the water is unusually exposed to the wind. However, in our book, *Tackle Making* (pp. 75-96) we do give a detailed account not only of how to make the main floats but also the circumstances in which each basic type is used. The assiduous reader can, therefore, take this question a little further, at which point he will discover that there are very few *basic* types of floats. So for small water fishing all you really *need* is a selection of Avons, antennae floats, wagglers and small quills. In addition, especially for rudd fishing, small bubble floats (or slow sinking wooden ledgers which do almost the same job) can be used. Even these can be dispensed with: a self-cocking Avon, appropriately weighted for the distance, does the job just as well.

Pole floats are slightly different and bring us naturally to the subject of pole fishing small waters. Neither of us knows a great deal about pole fishing, although we were blooded long before the modern pole fishing boom with its glass or carbon fibre poles, with shock absorbers on special floats. Buying a pole and its accoutrements is simplicity itself: with the help of a good match angler select the best you can afford. Make sure you can replace shock absorbers, and ensure that you have a wide selection of the specially designed floats. Our main concern here is *when* you would choose to pole fish on small waters. Clearly, overgrown streams and ponds are out of the question; but small rivers with good pools or long deepish glides are certainly possible; as are pond swims where overhead branches cause no problem; and small drains, particularly those with an easy flow, are well fished with poles. Although one can achieve wonders with the tiny hooks, fine lines and delicate floats, what you cannot do is seriously fish for big fish with pole tackle. So if you decide on a pole it is because you expect it to be a very efficient tool for removing small to medium weight fish to make up a decent bag.

If terminal rigs in pole fishing have become specialised, then they are flexibility itself when compared to rigs in the rest of coarse angling! What must be immediately apparent to the reader is that terminal rigs using 2 oz bombs are hardly likely to be necessary in small water fishing. Nevertheless, terminal rigs involving scaled down weights may be quite appropriate for putting ledger tackle on to a gravel bar at fifty yards range – whether you cast from one side of the pond or the other! The main problem with ledger rigs, and this does apply to their use anywhere, is that on the strike the ledger stop system may fail – be it lead shot or polythene tube and stopper. Apart from split shot there are a number of brands on the market which work on the polythene tube and stopper principle: they are certainly better than lead shot in that they are much less likely to slip, but they are not as good as the swivel system illustrated in Figures 19 and 20. Only two points need be considered here:

1. the stop system *must not* slip on the strike, otherwise the fish is not hooked;

2. three knots (Figure 19) are no weaker than one, despite common misconception, provided each is well tied.

We prefer the swivel system of Figures 19 and 20 to any other system.

The line from one swivel goes to the hook (Figure 19) and that from the other (or the same in *fixed* lead paternostering) goes to the 'lead'. We place the word lead in inverted commas because it need not be

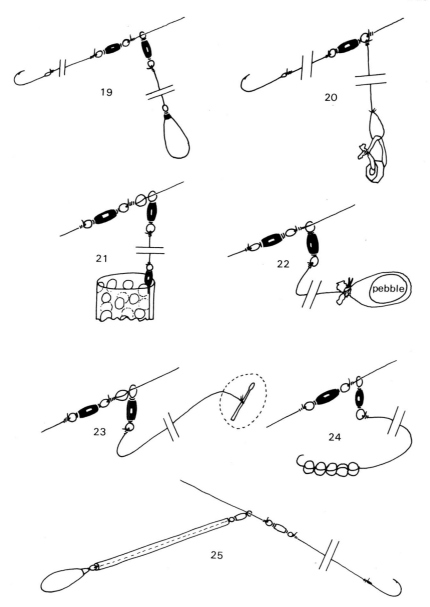

Figs. 19-25 Various terminal rigs based upon the preferred use of two swivels, one acting as a stop shot.

lead, of course: it can be a swim-feeder; a ball of mud attached to a match stick; an old nut attached with PVA; a string of split shot which will slip off the line when snagged; a polythene tube to stand proud of the weed, and so on. (Figures 21-25.) But the fundamental principle remains unaltered, to get a good, non-slip, weight-carrying paternoster or ledger system, as in Figure 19, and *then* vary the weight, the hook, and the two links to these, as dictated by the circumstances. For example, when small water barbel fishing one might well decide on a very long hook link and a small maggot-loaded swim feeder as a weight, but attached to a short link length of six inches. By contrast, when snatching small roach from the middle of a pond in deepish water, one might well opt for a small Arlesey bomb as the weight, on a three-foot link, with the hook link shorter at two feet or less and the line very tight, paternoster style, to the rod tip. Thus the crucial decisions in all these rigs are made early on; namely, choice of line strength, *our* decision to use swivels (which almost never fail), and only secondarily do you choose the 'lead' and hook appropriate to the circumstances.

Further special tackle is needed in pike fishing small waters. When writing *Fishing for Big Pike* Barrie expounded his successful principles of ledgering (free line or otherwise) for pike. Such methods were evolved in small waters, and the rigs given in the preceding paragraphs work extremely well except that for the single hook one changes, always, to a wire trace and (usually) treble hooks or hook. The reason why such methods work well, rather than the Fishing Gazette bungs mentioned earlier, is because on small waters pike are cautious: they know all about disturbance and the accompanying attitude of mind of the F.G. enthusiast. Big pike are not uncommon in small rivers and ponds, but they require a quiet and careful approach and appropriate tackle.

The terms long and short range are relative matters on small waters, but we would always opt to fish short range if possible so that there are unlikely to be big pike *between* you and the end tackle. By all means thump out a big deadbait, on heavy tackle, into the middle of a small pond, but you will almost certainly be fishing beyond the fish – or beyond some of them. Far better is it to fish with the lightest tackle you can (8-10 lb b.s. line if you are able) on free line or light ledger, and in a pre-baited hole close to the bank, say, up to thirty yards range. If, in time, this fails, then have a go at the less accessible water. Done the other way round, every retrieve of tackle disturbs the pike close in. We found many years ago that, particularly in shallow water, floats and lines going up vertically from the bait were off-putting to cautious pike. In short, one's whole attitude needs to be different when piking on small waters as opposed to big rivers and lakes.

Big pike will always take big baits on occasion and those in small ponds are no exception. But always remember the disturbance factor. On a small water you might just as well present a sardine or sprat, with chopped pieces of the same in a bread and bran groundbait. This gives you a much better chance of hooking on the strike, enabling you to strike earlier, and there really can be no doubt that the big pike will, before long, know that the bait is there! If one spot fails after a reasonable trial, then move on and try others. Eventually the whole water can be carefully explored and, even if the pike were not in feeding mood, at least they will remain unspooked as well as unhooked!

It can be appreciated then that both the tackle and the approach can be tailored to small waters; the tackle by, in effect, fining down and the approach by quietening down! Otherwise the rods, lines, and peripheral tackle need only be the same as in any other piking.

We have left until last the question of lure fishing for pike, simply because it extends the principles expounded above: everyone likes firing huge plugs to distant horizons, but on a small water you can put down every pike in the lake inside five minutes. Once again, use small lures and fine, dark, line before ever contemplating use of a nine-inch jointed Creek Chub Pikie. Stalking fish might be the order of the day, especially if that day is mild and sunny and you can see the sunbathing fish. A word of caution here. Should you be using wobbled sprat in preference to a small plug, then *wait* before striking a take. For some reason it is very easy to pull a sprat from a pike's jaws, without hooking it, even when you have watched it engulf the sprat. Resist an immediate strike, allow the fish to pull off three feet or so of line, and then let it have it with a firm pull rather than a powerful strike.

Stalking is a technique that can be used for other small water species of coarse fish, notably chub (usually in streams) and, *par excellence*, carp. The tackle need be no different to that we have outlined above – it will usually comprise free lining for example – but the baits may be changed frequently from worm to slug to wasp grub in chub fishing, or from floating crust to floating H.P. to slow slinking flake in carp fishing. In these, as in lure fishing for pike, watercraft is at a premium. And this last, perhaps the most crucial factor of all, we will deal with on its own in Chapter 15.

CHAPTER 14

GAME FISHING

Niceties of game fishing, things like a delicately fished dry fly delivered with grace, style and panache tend to 'fly' out of the window so to speak on small waters. Likewise, spinning becomes a matter of accurate casting, often to achieve a retrieve through, perhaps, eight or ten feet of water, and even the humble worm that is trundled downstream requires a sixth, almost radar type sense on the part of the angler if he is to accurately place its progress among pools, falls and snags that always abound.

Confidence is the key to successful – and happy – small water game fishing, plus an initiative on the part of the angler which will lead him to adopt or adapt any of the styles where and when the going dictates. Rigid adherence to accepted methods or even minor details such as the position of a hand on the rod, the make of a line or the colour of a pair of waders can become the path to ulcers. Remember the old saying – Man who keeps his hook in the water the most, catches most fish.

FLY FISHING

TACTICS

Artificial flies catch fish. It doesn't matter two hoots whether the fly is presented wet, dry, damp or even, at times, whether is is suspended an inch or so above the water. If it can convince a fish that it is an item of food, is apparently unconnected with humans and it does not shoot out of sight when interest is aroused, then the chances are that it may be accepted.

All of which makes for sport and should encourage those who feel some embarrassment that their casting and general skills may not be quite good enough to get up and have a go. We have both fished just about every type of water imaginable and we are decided that when it comes to a style of fly fishing we prefer, if possible, to fish upstream, using traditional dry fly tackle and a fly that is a reasonable imitation of the natural, which does not necessarily have to be on the water at the time we are fishing.

But we are equally at home fishing the dry fly downstream, throwing a slack and snaked line across and slightly down, allowing the fly to drift naturally under overhanging banks and bushes impossible to reach by normal casting. Of course, the wet fly can be fished in the same way – up or down – and we rarely bother to change to a sinking line for the style, continuing with the dry line but sinking the leader by means of one of the proprietary liquids now on the market. The result is a sink tip which has the dual advantages of saving money and providing a focal point where the leader disappears under the water, helping the delicate decision on when to strike. It is a combination that is ideal for the most difficult of all fly fishing styles – nymphing. On those occcasions when there is not enough water to enable an artificial nymph to swim we switch to a monofilament line, but more of that later.

Accuracy is everything when casting on small streams and we have found the principal ingredients for success have little to do with style. First of all, you should keep a short line. Easy to say, we know, but not always easy to obey especially where one is keeping off the skyline, back from the water, and avoiding obstructions that may be in the foreground. But if you study the water you want to cover before waving a rod you will sometimes see an approach that can overcome some of those problems and, if time is taken and a bull-at-a-gate approach avoided, a short accurate cast can often be achieved.

The forward tapered line as an aid to accurate casting is something we mentioned earlier, but we repeat our observations again – where short distance casting is the order of the day then the forward taper will work the rod long before a double variety, effort will be reduced, and accuracy improved by as much as 100 percent.

Finally, there is the question of length of leader. We see people struggling with leaders (casts, call them what you will) which are often longer than the rod itself. Thinking behind this appears to be that the further there is between heavy fly line and the fly itself, the less chance there will be of a fish being frightened. This is a problem of contradiction. Long, soft leaders do not turn over well, seldom land where you want them to and inevitably force you into making more than one cast to the fish – provided, of course, that it is still there.

Ken pondered this question of leader length and suppleness and finally threw it into the lap of an expert, Jim Mackenzie-Phillips, who runs a fishing tackle emporium near Northallerton. His

recommendation of the Berkley Big Butt leader, and the accompanying instructions of how much to cut off, and the precise length to add at the tip resulted in a short, firm, nicely tapered and manageable leader, easing and assisting accurate casting.

We are often amazed at the number of fly fishermen who use the traditional forehand cast morning, noon and night without regard to water or bank conditions. Indeed, we have seen anglers walk away from rising fish in a difficult but not impossible position because they felt they could not get a fly to it – a pretty accurate decision in most cases with only the single style at the anglers' disposal. The stalemate could probably have been broken either by casting from the backhand or by using a horizontal cast.

Neither of these casts are difficult. When casting backhand, grip the rod in exactly the same way as you would for a forehand cast, but throw the rod back over your left shoulder as opposed to the normal, right shoulder (*vice versa* if you are left-handed, of course) and turn the wrist so that the back of the hand is uppermost.

The horizontal cast is more difficult to master – at least, we found it so. Here the rod is held low over the water and line is worked horizontally forwards and backwards over the surface. All is usually well in this preliminary stage, and some distance is achieved. The crunch comes when the cast ends – and all too often it comes with a splash. Remember that, as with the forehand cast, some loose line should be released at the last second. To allow this to shoot clear through the rod rings turn the wrist over when releasing the spare line and this should bring it into a reasonable roll to the surface without too much splash.

There are bound to be as many abortive casts, with the fly hung up in trees or on bushes, as there are successful ones in this type of work. The sensible angler does not snatch and stab with his rod to free the hook, but instead reels in loose line and applies a steady and increasing straight pull until the line works free. Yes, the fly may be missing and yes, the leader may be strained. But with a little repair work a fresh cast may be made to the self-same fish in minutes, as opposed to some empty water facing one after a snatching-and-heaving match.

With badly overgrown banks one is usually forced to wade. Before plunging in, fish the water around you as far as possible first – the number of fish lying under the banks you are standing on may be just as many as those you hope to encounter under the undergrowth you will be struggling to fish, and far easier to reach.

Heavily obstructed waters and those that are gin-clear will usually be avoided by even the most serious fly angler. But they can be successfully fished by a fine monofilament line, with a modified paternoster rig using one or two swan or BB shot pinched on the end of the line, and one or two short links leading outwards above them, the flies being attached to the ends of each of these. A fixed spool reel allows reasonable casting with this featherweight rig and the flies can be worked upstream or down, the lead being allowed to drag, or even pumped mackerel-feather fashion, in a sink and draw action.

As if you have not already guessed, this is also an ideal medium for presenting a nymph, using one link of fine nylon that will 'flow' away from the main body of the line and its terminal lead.

The natural extension to this form of fishing is a dapping rig. Many anglers know the principle but few fish it correctly. The idea is to use a pierced bullet held by a plug or stop shot some 12 inches or so above the fly/slug/grasshopper, etc. bait. The rod is then eased between branches and bushes, extended outwards over the surface and line is gently unreeled to carry the bait down into a gentle kiss with the water's surface. It is a deadly method ruined by many who omit the lead, wind line around the rod tip and then unroll this when the rod is in place. Usually the line is not long enough to reach the surface, so more line is freed and forced through the rings by jigging the rod up and down – a movement fatal to surprise and guaranteed to frighten good fish away. This form of dapping is not to be confused with extra long rod dapping, widely practised on the big loughs of Ireland, and increasingly so on big reservoirs in the U.K.

We cannot leave this outline of tactics without mentioning night fishing. Few seem prepared to try it on small waters – the average excuse being that it is difficult enough in broad daylight. But night work is not impossible. Study the course, work out the best vantage points – where you can cast from and where a fish can be landed. Mark both places with white paper, staked to the ground or fastened to a convenient tree or bush.

Once you know where to cast, sit and work out a safe casting distance for the water. This will be somewhere between the maximum and minimum length, and mark that distance on your line by two or three turns of knitting wool. That marker will tell you that the safe distance has been reached as line feeds through your hand at night – and will cut down snagging drastically.

THE FLY

In recent years the artificial fly and the way in which it is tied seems to have assumed greater importance than the way in which it is used. It would almost seem it has to deceive the angler that it is a perfect reproduction of the natural insect, let alone the fish. Any that are found to be tied with a turn too many on the body or a whisk too many in the tail, are often cast aside as useless.

Perfect imitation may be necessary for those who fish the southern chalk streams, where fish have time to minutely examine every offering that passes, rejecting frauds immediately. On a small stream, especially one overgrown or where the current moves at reasonable pace, fish are unlikely to expect anything other than a natural item of food – food which may not be plentiful. So in nine cases out of ten a detailed inspection is skipped and vague look-alikes are jumped on.

Of course, there are anglers who make no attempt to imitate a natural fly and fish largely with attractors; gaudy efforts for the most part that probably look more like a small fish or similar aggressive underwater insect. Which camp you subscribe to – imitator or attractor – is a matter of personal choice, though we feel that it is better not to get too bogged down in the strict imitator brigade; if you follow this line to its natural conclusion then a bottomless fly box would be required for each small variation, every tiny insect – and more time would be spent tying and changing flies than actually fishing.

We attach more importance to size than accurate colour and shape reproduction. An obvious conclusion, of course, when you consider a small fly on a river heavy and coloured with recent rain or a large fly on a river summer-thin and gin clear. Time and again though we have fished a particular fly under perfect conditions without success, only to meet with fish the moment we changed to a smaller or larger version of the same pattern – which may have become so bedraggled as to bear little resemblance to the pattern in which it was tied. Like the song, it appears that 'It ain't what you do, it's the way that you do it'.

Winged or hackled? Heavy or sparse dressing? Thick or thin wired hooks? We don't think that it matters a jot. Providing a fly floats well and sits on the surface then it is a dry fly. If it does not, then it's in the wet category. We prefer our wet flies to be tied with soft hackles so that they make a lively movement, and where there is a ribbing of silver or gold we like it to be visible and not tucked under folds of body material. Other than that we tend to keep a range of sizes rather than patterns and, perhaps, it is as well if we clear our yardarm at this point and declare them, together with some reasons for their preference.

Caddis larvae we find to be plentiful all over the country and a representation from the sedge family is a 'must'. The red sedge fits our bill and we also like the Welshman's Button that often fishes well for sewin (sea trout) in Wales, taking fish when the red sedge has been refused. At the other end of the scale, small midges and gnats are invariably present on every water and fish feed on these small insects with ferocity at times. Impossible to imitate successfully, even when tied on size 22 and 24 hooks (they pull straight out of the mouth on a decent sized fish), we stick to the Small Black Gnat and keep the faith when fishing it. It is bound to take at least once in an evening, and often sorts out the men from the boys as regards size. Ken's best fish was a 5 lb sewin on a size 16 fly from a stream not ten feet wide!

Hardly a month of the year passes without a hatch of olives. We keep both light and dark varieties in our box, with the odd Lunn's Particular as a choice for the spinner. This, incidentally, also pulls in the odd fish when the 'smut' is on the water. For late evening and night fishing we would not be without the Coachman, which is as good a representation of moths and night bugs in general as one can find. We don't worry too much whether it swims or sinks; in fact, we often welcome it kicking out a little disturbance if it catches and skitters across the surface; it's an action that bring trout, sea trout, chub and even perch to take with a bang.

One wet fly we reserve for waters with a good head of freshwater shrimps, the Gold Ribbed Hare's Ear, both with and without the weighted body and likewise the Pheasant Tail Nymph which we use with lead for pools, without lead for the short runs and shallow reaches. Finally, two attractors we are never without, Teal, Silver and Blue, which bear a striking resemblance to small fry of every description and the Alexander, which resembles God knows what, but returns fish for us from stream or still water without fail in the course of a year.

Of late there has been a spate of imitations on the market along the lines of back-swimmers, beetles, bugs and other crawlies. We have found but little use for them – or to be more honest, we have found that results from our normal fly range has been sufficient to keep us from dabbling too much into the unknown. One 'weirdie' only occupies a corner in our fly box, and that is the Crane Fly. Allowed to drag, half drowning, through a pool or under an overhanging bank it is a winner, especially around the mid-afternoon lull time when little appears to be on the move. (See Photo 64 on page 121).

The above choice holds good for small still waters as well as streams and small rivers. Yes, there are a dozen or more excellent streamers and general reservoir lures that all catch stillwater fish. But before you rush out and buy them, fish the water with a selection of traditional wet flies first. Then, if you are without success, consider them by all means. Personally, we don't think you will find them necessary.

SPINNING

The vast majority of game fishermen regard spinning as a task to be undertaken when the water is too high and coloured for fly fishing. In fact most of their tackle emphasises that, with heavy rods, thick lines and heavy, unbalanced fixed spool reels as the norm along the banks. And in nine cases out of ten you can lay odds that the lure being fished without regard to water conditions is a large and heavy Toby type.

Spinning is a way to present an artificial *or natural* lure to a fish in such a way that it looks like a super-de-luxe meal. But it follows that the most killing lure will be wasted unless it is fished in all the right places and not thrown out, chuck and chance it, along or around a small water. Awkward holes and corners where one stands to be snagged, and where there may only be five or six feet of clear water through which the lure could be made to work, must be acknowledged as an everyday challenge for the small water man. It takes concentration and first-class tackle to succeed.

LURES

Those who fish small waters must always remember that heavy lures, though easy to cast, take time before they begin to work efficiently, and time during a spinning retrieve represents distance. At the other end of the scale light baits work quickly on the retrieve but seldom fish deeply, and are apt to cut across the surface out of sight and hearing of many fish.

A compromise? In a word, no. But there are more methods than one of spinning and baits worked sink-and-draw or used in conjunction with a modified paternoster rig are killing styles that bring results and certainly deserve more popularity and support than they do at present. Those rigs, and many of the lures that will be needed to replace them through loss by snagging, can be home-made. What artificials can and cannot do and which can be made by the angler are best summarised as follows.

Devons: Not very brilliant as attractors. The light pattern emitted during retrieve is not great and they are slow off the mark to revolve at all unless there is a reasonable current or good distance which will allow a faster than average retrieve. They are ideal for pools, especially deep ones where the trace can be weighted to give extra-quick sinking if required. Wooden models are also fine for long shallow rapids and glides where heavier lures would be lost. Several firms, especially

McHardy of Carlisle, offer wooden or plastic kits that can be home assembled and painted (they even have slightly imperfect models for sale at times – ideal for cutting cost on heavily obstructed waters). It is worth remembering at all times that action from a Devon is only as good as (a) the bead or block against which it revolves and (b) the state of the swivel at the head of the lure.

Quill Minnows: including the heavy but delicately coloured Irish Minnow. These lures are so natural looking when retrieved that they always do well. They are similar to the Devon with the light pattern they emit, though the old maxim of 'bright for dull days, dull for bright days' is worth keeping in mind. They can be made at home – we describe how in our book *Fishing Tackle* – but it is a slow and tedious task. There is also a plastic quill minnow on the market. (They'll offer a plastic pop-up picture of Isaac Walton next!) Cheap, yes, but nowhere near as killing a bait as the real thing.

Use the Irishman in deep pools and leave the quill for normal stretches, preferably where there is a fair length of water through which a retrieve can be effected.

Bar Spoons: Excellent for most water conditions. They are available with a range of weighted bodies that enable deep water to be covered and possess that great advantage over most other lures in that they can be felt working through pressure transferred through the rod tip. One knows on the instant when weed or similar snags have been touched, or when the lure has ceased to work properly. The light pattern emitted is excellent, there are innumerable permutations of colour available, and whilst not cheap, they don't completely break the bank if two or three are lost on an outing.

Disavantages? Yes, a tendency to 'lift' or dodge sideways when cast into a strong wind, leading to some inaccurate casts at times. They are now also unobtainable in kit form, which means some rather tedious processes for the home manufacturer.

Spoons: The original of the species, tea-shaped, together with Toby, Salamander, Norwegian, Jim Vincent – there are a double handful of varieties all of which are cheap, besides being ultra effective and capable of working in most extremes of water, either still or moving. Easiest of all for home manufacture, either from sheet metal or the domestic utensil, they can be painted or polished in most combinations. We make them our first call for new and untried waters, often working them sink-and-draw without the addition of lead on the line.

Natural Baits: Minnows, Gudgeon, Bleak, Sprats – each and all are excellent and cheap baits which, if not actually spun by means of a vaned mount, can be worked sink-and-draw or just plain wobbled through almost every type of bankside hang-up imaginable. If they snag, little of value is lost. And they do fetch fish – big ones, especially sea-trout and back-end salmon that are working their way into the opening of small streams leading from the main river.

Kept wrapped in paper strips, ten to the pack, then frozen, enough for a day can be collected in seconds, carried in a small plastic box and mounted as required. Small baits, minnows etc. worked on the deadly 'drop' principle can be used with a single hook and lead rig. Bigger baits can be used with spinning mounts that are sold by all good-sized tackle shops. Preserving, mounting, mounts, colour and colour dying we dealt with in our *Spinners, Spoons and Wobbled Baits* book.

TACTICS

Swivels, anti-kink devices, traces, hooks and weights – all these we presume you know, together with where and how they should be used. We merely add a caution that because you are fishing a small water where you may lose tackle, do not use cheap accessories. At the end of the day they usually become the most expensive by virtue of the quantity you lose!

Small water spinning differs from all other forms in one essential – you will get one, and only one chance of casting to a fish, or through an area where a fish may be. Either you connect or disturb – second casts are rarely needed or successful. That single chance means that the lure, whatever it is, must be:
1. acceptable
2. accurately placed, and
3. correctly fished.

Acceptable means what it says – acceptable to the fish. It should not be the self-same lure that has been unsuccessfully chucked around for the past hour or so. A deep pool needs a heavy lure, shallows need lightweights that are less likely to snag. Previous outings may suggest that a particular colour or shape is best, in which case care should be taken to alternate sizes through the day until the right one is found. Add to that the water colour, light factor (hard, soft etc.) and the rate of retrieve needed to make the bait effective and you will recognise that acceptable lures are the result of considerable thought, not just luck with what you pick up from the tackle box.

Accurately placed lures come from outfits correctly balanced from lure back to reel, with which the angler is completely familiar. It may sound boring, but when did you last practice accuracy casting away from the water? Practice on the day at the place where you are hoping to make contact with a fish can only guarantee failure. Knowledge of the water, places from where a cast can be made without disturbance and where the line can be kept short are sure-fire winners; remember that short casts mean a short line, which ensures accuracy and control.

Correctly fished, it will not matter whether you spin a lure upstream or down. What will matter will be the rate of retrieve, which is infinitely more difficult when bringing a spinner down through fast flowing water. Counting the bait down as it sinks so that each cast encounters a fresh level of water, varying the rate of retrieve and the position of the rod tip to alter course and break monotony are natural actions to the skilled spinning man.

Fishing a paternoster rig where the terminal weight will hold position in the stream and allow the spinner to work from a long trace is a killing method but it requires discipline. The tendency is to use too large a lead, dragging the bottom and disturbing it. The secret is to use the smallest lead practicable and bounce it slowly back to you. Most anglers spin too quickly; if you fall into that category then use a bar spoon which will tell you when the blade stops turning through the vibrations on the rod tip. Remember also that the position of the bait in the water is decided by the length of line between weight and junction with the trace. It can be varied!

PLUGS AND PLUGGING

Another sad case of neglect on the part of the game fisherman. The plug can be as deadly as the spinner – more so at times for the plug fished on the surface at night. A small floater which need not be magnificently shaped or coloured in any other than something light, is medicine for most fish – sea trout in particular.

The stimulating thing about plug fishing is that by carefully selecting the correct lure you can cover every eventuality. Floaters can be worked willy-nilly with little fear of snagging, whilst floating-divers may be allowed to drift below the place where one wishes to show the lure, then worked back again. A snag in the water can be 'skipped' in much the same way – float or cast out, retrieve, stop winding before the snag and allow the lure to surface, tow gently over the obstruction and then fish on.

With one or two notable exceptions the plug is a dead duck on most tackle shop shelves. Those available are often in very large sizes, and floating/divers. Fortunately, plugs are easy to manufacture at home and inch to three-inch floating types with blunt nose, shaped from wood and supplied with a treble to scale, make excellent lures – especially evening and night attractors.

In the same category are poppers, the American small plugs with great rafts of whiskers and rubber legs fixed to the body that stick out like hay stacks. Several firms stock them and with the aid of a little trimming these make first-rate lures for both trout and sea trout.

Mention earlier of night fishing merits a further paragraph. One can spin with great success at night – or rather, into the early hours – after which the spinner, in our experience at least, seems to lose its attraction. Highly reflective surfaces on the various lures are not strictly necessary, most fish feeling rather than sighting movement. Again, spy out the land and work out an attack in advance, marking danger areas and places where a fish may be landed.

WORMING

Every seasoned angler will admit that this is not the easy task that it sounds. To produce fish when the water is in spate or with plenty of colour on is one thing; to even approach fish when the level drops and each stone can be seen on the bottom is another.

Barrie remembers climbing five hundred feet once, almost straight up and including more than one waterfall, in search of hill brown trout, only to find that he couldn't get near a single pool from the downstream end without frightening fish. Casting from thirty yards away into an eight foot wide pool, which itself is possibly at a level five or six feet higher up, is not easy, but to go any closer meant that the fish seemed to detect him in some way even when he thought he was well below the skyline.

In the end he solved matters to some extent by freelining from upstream and floating the line down some thirty yards (including several small falls). Snagging up behind rocks was very easy and hitting the bites most difficult.

Barrie also kept a detailed account of returns from a hill brown trout fishery and these are summarised in Figures 26 and 27, one a main pool on the stream, the second a spate pool only. What is clear is that in the undercut section the better brownies were on the bottom and

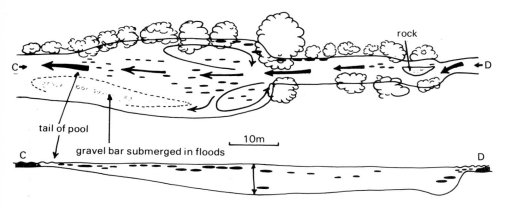

Fig. 26   Plan and longitudinal section of a north-western trout stream. In the upper diagram the relative size and daytime position of sea trout is shown; also shown, in the lower diagram, is the night-time position of sea trout.

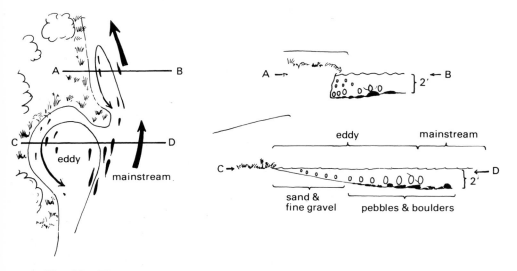

Fig. 27   Flood eddy in plan and section on a small trout water. Position and relative size of fish are shown.

close to the bank, whereas on the shelving bottom (section C - D) the bigger fish were always on the edge of the main current. Out of spate conditions few fish were found outside the pool of Figure 26, where they took up the size distribution shown.

Sea trout do not behave in the same way at all, the bigger fish being towards the tail of the main pool most of the time, except at night when they could be found in what remained of the pool in Figure 27. Mind you, we are talking here about brown trout averaging $\frac{1}{2}$ pound or so, and sea trout around the 1 pound mark.

Fine water conditions demand the longest rod manageable on small waters. With a 5 ft rod you can't hold much line off the water when worming and this is often vital. Line size will depend on the water – but we have never exceeded 6 lbs and that on waters where the odd sea trout or salmon may appear. We find the nylon link, wrapped over the line and holding such shot as necessary, to be the best where snags are prominent – you can pull the leads off without making a break, and as there are no swans in the mountain streams we have in mind no-one can get uptight about it.

If you have a lot of walking and climbing to do we suggest you use an elastic band to hold the tip and butt of the rod together where it needs to be dismantled and carried – a sensible precaution that has saved us from breakages where the going is tough.

Float fishing the mountain streams can really only be done if you have a long, deepish pool. If you get the chance to do this, then feeding maggots (where allowed) can be most effective. Clear water can be fished near dusk and a white vane added to the float tip will improve visibility. As far as white water is concerned we have fished it effectively with a float, but not *more* effectively than by the upstream or downstream freelined worm.

One technique which can be very successful is either a maggot, caster, grasshopper, slug or worm fished on the surface or just below it, with a bubble float a little up the line. All of it a bit Non-U perhaps, but effective at times.

In summary; mountain stream fishing is to lowland fishing as the rapier is to the cutlass; attempts to be violent and crude fail, but intelligent thrusts here and there often succeed – lifting this fishing into a category of its own.

CHAPTER 15

WATERCRAFT

Let us begin by telling you what not to do: a true incident which happened only a few years ago on a small tench water in Yorkshire. At 6.30 a.m., long after a beautiful, quiet and misty dawn, a furniture van drove up on to the bank. It didn't stop three hundred yards along the lane, as it could well have done, but forced its way along the rest of the leaf-hung lane, finally turning through a gateway and up on to the bank itself. Thereupon about a dozen very rowdy anglers spilled from the back, after the tailboard had hit the ground with such force that the vibration could be felt fifty yards away. They proceeded to circle the lake – about sixty yards across – finally clumped into all the available swims, and tackled up with a great clattering of boxes, each and every man standing bolt upright within three feet of the water. They didn't get a bite between them, until by 7.45 a.m. they'd had enough, pronounced the water a total write off and the whole noisy rigmarole was re-enacted in reverse.

Had the lake in question been really big they might just have got away with it; after all, many anglers behave like this. But on a tiny, shallow lake, even the small perch went off the feed. One's approach to small waters has to be extremely careful – every facet of it from clothing to tackle.

In the first place a longish walk to the waterside is not a bad idea unless the lake is near a road and the fish are used to cars. Before actually walking up to the water it really is a good idea to stop at, say, fifty yards distance and talk amongst yourselves, or if on your own, whistle a happy tune or something! This has the effect of making the water birds paddle quietly for the shoreline and the safety of reed beds. Just as animals are warned by the call of a disturbed blackbird so are fish sensitive to frightened water fowl leaving the lake surface in a flurry of spray. With some waters, mountain streams for example, such behaviour will not be necessary, nor will it if other anglers are already on the lake.

Frankly, we cannot imagine any angler being too quiet. Remember that sound travels nine times as fast through water as it does through air: put another way, the same noise sounds louder at the same distance away from the source if it is received in water. Much more commonly one reads of advice about keeping below the skyline. This too cannot be over-emphasised. Even in a murky, flooded trout stream the trout is quite well aware when the sky is blacked out by a human figure. In clear water they seem to take less note of cows, with which they are presumably quite familiar, than they do of the human silhouette! The trout's 'window', equally applicable to other species, has been quoted many times but is nonetheless valid: if a trout is six feet below a calm surface it can see clearly through a window twelve feet in diameter. But objects outside this window for a distance down to 10° above the horizontal (usually the bank) *are* visible as part of a fuzzy skyline. As it swims shallower, the window gets smaller and its chances of spotting you in its *clear* window recedes. However, we are less happy about the commonly held view that choppy water also reduces the trout's awareness. It may make the picture more fuzzy, but it *enlarges* the window for a given depth and surely a silhouette will be just as visible in its totality if not in its detail. To put it briefly and bluntly, get as low down as you can, consistent with reasonable comfort. On small lakes this is usually no problem, but on a fenland drain it may mean crawling, as it may well on a mountain stream.

Lowland 'trout' streams are much easier by comparison because you often have a deal of lush cover. Here the real problem can be the vibrations of boggy margins and the vibrations carried along tree root systems. Anyone who doubts the latter should get a friend to climb to the top of a tall tree and scratch the bark gently with a finger nail. The sound can be heard in the trunk by putting your ear to it, or even well along the spreading roots where these touch the surface. So avoid tapping out your pipe on the bole of the nearest tree!

All these factors are relevant on other types of waters, but on *small* waters everything has to be that much better attuned on your part. Even clothing is important; it should be drab and dull. In times past this used to be quite natural because all anglers went fishing in old clothes, tweed jackets, sweaters, old cord trousers. Nowadays you see the most garish sweatshirts, with slogans in bright colours, as well as loud sweaters with *fish* on them. Quite apart from any aesthetic appreciation, avoid them because they are fish scarers, as are shiny rods.

All tackle in use should be dark coloured with a dull, matt finish. One deplores the modern brilliantly shiny and painted reels and thickly varnished rods. A flashing rod can be seen up to a mile away. Fish have no trouble whatever seeing them in the fifty yards range we are discussing in this book.

Of course, we should have less difficulty convincing you if we lived in the days when stories of the American backwoods were legion, or, indeed, when Boy Scouts roamed the countryside in packs! It was part of the code to behave as quietly and inconspicuously as possible and to do it with pride and skill. But that is exactly the attitude that is needed for watercraft on small waters.

Having got the approach work right and having moved into position quietly and almost invisibly, you have to begin fishing. And here it is possible to undo all your previous skill by, for example, heaving into the water huge, solid balls of groundbait. In small waters we have noticed that fish are wary of swimming over a bottom coated in white groundbait: so use dark coloured groundbaits. Avoid the use of heavier leads and floats than is necessary to get the distance needed. These rules apply particularly on shallow waters, but everywhere to varying degrees.

Personally, we also go for dull or black lines, matt finished, dark nets, floats with the very minimum of bright colour and dully-painted leads. As you can imagine, it's really a question of developing an attitude of mind and application of a little commonsense. If the fish cannot see you or hear you they will not be frightened. Q.E.D. Any other attitude or approach is difficult to justify.

CHAPTER 16

MEMORABLE DAYS

It was thirty yards across, barely three feet at the deepest point and bespattered with yellow lilies, lined with flags and part-choked with leaves from horse chestnut trees close by. As ponds go, most people would pass it by and not give it a first, let alone second glance. Damsel and dragonflies knew it, so did the moorhens that nested in a small clump of trees that had somehow sprouted and now flourished on a shallow ledge towards the water's centre.

At five in the morning dew soaked both shoes and trousers, leaving them cold and clammy against the legs. Later, summer's sun would dry them, stiff and unyielding to itch and chafe, but at that early hour one just shivered with the cold – and anticipation. Between the lilies, beyond the flags there were needle point bubbles rising in fine clouds, like fizzy lemonade poured into a glass and left to stand.

Gently, inch by inch, the red goosequill float was lowered into the midst of a spectacular path of bubbles. Gradually the line was eased downwards until lead weights cocked the float, leaving it to turn slightly, first right, then left before it centred and stopped, a beacon that glowed among the yellows, greens and sombre browns around it.

Resting on the bottom was the bait, made from flour and water, moulded until it was like putty and could be stretched out into strands, then rolled back into the ball. But this was no ordinary flour and water paste mix. To it had been added a dessertspoon full of sugar, and another of custard powder to colour, sweeten and, hopefully, attract every passing fish.

Two early doves crooned; the moorhens, heads bobbing, searched around the opposite side of the pond for insects and weed strands. Slowly the light increased its strength and bubbles came less frequently. The nagging, growing doubt that it could be all over, that the fish would disappear below the lily pads not to feed again until the morrow, became an obsession.

The first movement from the float was small, a gentle swing which, for several heart-stopping seconds gave the impression of imagination until a second swoop, a gentle dip that sent three small ripples away from the float outwards towards the lilies persuaded one otherwise. Anxious fingers gripped the rod so tightly that the tip quivered. As the float dived under the wrist and forearm jerked up, sending a shockwave along the rod's length.

For a second, even two, there was no sign of movement. Then the water boiled as the fish below was pulled to the surface. No quarter was given, no line released and in one hard, savage action the deep olive green and golden body was dragged to the shallows where it was scooped onto the bank with mud, weed and a tangle of line. By rights it should have escaped – the rod, a garden cane, was incapable of letting out the sewing cotton that served as line, wound around its end, whilst the bent pin used for a hook had no barb to prevent loss.

Measured against the handle of the cane which later was checked against a sewing tape, that fish registered nine inches. Not a record-breaking monster by any standards but when the fish is the first you have caught, and you are just ten years old, the memory of a first tench lasts forever.

* * * * * * * * * * * * * * * *

When small waters have played a large part in one's life then it is hard to recall one particular day that has been made memorable above all the others. Most slide together and blur into a pleasant background. If one wants to concentrate – say, on a big bag of decent fish – then I suppose there was the day of the pike.

With hindsight it should have been obvious that it was going to be a good day. There had been three weeks of steady, drizzling rain that had filled the ditches, brimmed the brooks, lifted small streams to their maximum and filled the main rivers around a thirty mile radius with muddy brown water. All in all, a typical late winter picture that delayed my setting out until mid-day.

Because of the colour on the main river – not, note, because of the current that was running – a decision was made to fish the stream just a mile above its junction where it was barely seventeen feet across at the most, and considerably less than that for most of the distance available for angling.

It was at the time in my fishing career when I did not really regard small waters as being worthy of anything great in the way of plans or style. Into my bag went some deadbait mounts, single size 8 trebles tied to cabled Alasticum, an eight foot spinning rod and matching reel, and a single pack of ten frozen sprats. As an afterthought a solitary spoon, copper and silver, was thrust into a side pocket before tracks were made.

It rained. Not that it mattered – wind destroys my angling enjoyment with its constant line-tangling and noise. Rain breaks the surface stillness and provides cover without annoyance. Perhaps it was this disturbance that brought some bearing on the day, for no sooner had the sprat hit the water when it was away; in fact, I had not really made a cast as such but thrown out the line to ensure that the float was free running and would hold against the stop bead.

There was no great quality in the thing, as my Irish mother-in-law would say; eight pounds of fish that fought well, looked good and shot away from the bank when it was returned. A fresh sprat, another cast – and away went the float to return with fourteen pounds of pike under it, firmly fastened to the single treble.

I began to have ideas about the day and sure enough, the next cast continued with the run of pike – and the next, and again after that until I realised I was in for a great day, but out for the count when baits ran dry (it was 25 miles for the home run to replenish the bait box). From then on I played the dead sprat and not the pike. To prevent pouching or even holding beyond the first ridge of teeth, I began pulling as opposed to striking and this lost me several good specimens that simply let go when they reached the bank.

The final score to dead baits stood at 14 fish for 10 sprats – not bad, but I felt like apologising to the last pike whose considerable age and $18\frac{1}{2}$ pounds weight should never have been insulted by being offered a sprat that was gutless, tailless and badly chewed. But sport didn't cease there. The spoon accounted for a further three pike, or rather jacks, all under four pounds and quite happy to take up to, and after, darkness had fallen.

Of course, I was there the next day but didn't get a touch. Mind you, the water level was down, heavy currents on the main river had caused a bank to sag, then hole, and there was nearly a mile of flooding across the marshes close by. And the day of days has never been repeated at that spot. I reckon I had hit it just right, when coloured heavy water in the river had sent fish of all sorts into the sheltered tributaries. Certainly they would not have made that particular stream if I had not cleaned its entrance of reed and silt during the preceding summer. Whether the pike were actually in search of shelter or food I just don't know, but it was, for me, a memorable day that made a happy man very old – if you know what I mean!

* * * * * * * * * * * * * * * *

A small stillwater in the East Riding of Yorkshire provided some real lessons in angling for Barrie. It was, at most, fifteen yards wide and, perhaps, seventy yards long, an old borrow pit adjacent to a railway embankment. The depth was an average of four feet but no deeper than seven feet anywhere. Both banks were densely lined with hawthorns, and slit-like swims had to be cut through the hawthorns to get a cast in. Fishing from the ends of the pond was impossible; unfortunately, because a much better coverage of the water would have been possible. As far as is known this water still contains only roach and eels. Both species were reputed to be small: both, in fact, grow big. The only other factor Barrie had to contend with was water moss (*Fontinalis*) in profusion. Occasional draggings produced huge bunches with millions upon millions of contained freshwater shrimps (*Gammarus*) and water louse (*Asellus*). The hawthorns themselves overhung the water and in several places one could lay on and between water moss clumps, and almost underneath the overhanging branches.

It is worth remembering that whilst one becomes very skilful at casting on small waters, learning to avoid overhanging twigs with arrogance, in wet weather the branches hang lower. This simply means that skilful casting with the sun shining needs to be modified. An overhanging willow will droop at least a foot lower when the foliage is wet. It was just so on this morning of steady drizzle that Barrie found himself in a narrow swim between two bushes. After losing two size 14's to gut – an expensive morning by any of the (then) standards – he got the hang of it.

In those days he had neither brolly nor basket, so sat on a wet bank on his rolled up holdall (home-made, of all things, out of an old raincoat) with an ordinary umbrella (minus two spokes) clutched in his left hand. The swim enabled him to fish laying on under a trailing hawthorn on the far bank in about five feet of water. Each fish, usually roach of 2-7 oz, had to be brought in across a mass of water moss just in front of him. Twice he hooked, and landed, good eels of 1 lb 14 oz and 2 lb 2 oz which gave quite a battle on the $2\frac{1}{2}$ lbs line. They went into the keepnet with the hooks in their mouths. Four hooks down and only two to go. At about 10 a.m., just as the steady drizzle began to seep down his neck, the float dropped to its very tip and held – the perfect roach bite – and the strike, expecting another roach, got a solid shock. No eel this because it ran between the water moss beds. When it eventually slid to the bank (no landing net) Barrie could not believe his eyes – a huge, fat, deep roach which was weighed in eventually at 2 lb 5 oz.

Having no balance he 'sacked' the fish in his raincoat holdall, packed up and cycled ten miles to get one, and a camera. Although

79. The actual swim from which Barrie took his 2 lbs 5 oz. roach.

178

seemingly sacked up quite efficiently (sacking was almost unheard of in those dim and distant days) the fish died within two hours. Later inspection showed it to be an old fish, but very robust looking, and certainly a true roach. It remained Barrie's best roach until 1983 and his only two pounder, subsequently joined by eleven others at the time of writing. But the ecstasy of that first one ...

It was not altogether a fluky capture as he had been visiting the water regularly and had been separating the good roach to 7 oz from a big shoal of fry by feeding the latter cloudbait in the shallow water of the next swim. By this technique the fry stayed in one corner and the eels came along at intervals to investigate, whilst the roach angler continued to take good fish next door. Later, Barrie had several other good fish including some over 1½ lbs. And some fell to fly-and-bait fished slow sinking under the overhanging hawthorn bush. So, memorable days and good fishing on the tiniest of ponds. This pond is still there, and, perhaps, we'll give it a try again. Nobody fishes it today.

CHAPTER 17

AN A-Z OF MANAGEMENT PROBLEMS

ABSTRACTION

Taking of water from any surface or groundwater source is controlled by the Water Resources Act, 1963. The need for a licence is determined by many factors which include the type of source, the contiguity of the land, the use to which water is put and the method of impounding, etc.

If a licence is required for a particular situation then the applicant must advertise it in a local paper – and the *London Gazette* – and the proposal must be available for public inspection. Objections are taken into consideration with comments from the Water Authority if – or when – a licence is issued. Ken is deeply indebted to Mrs J. R. Frost of the Welsh Water Authority for much of the above information – she succeeded in explaining where trying to plough through the wording of the Act itself produced only instant sleep.

ACCIDENTS

One tends to think they can never happen whilst doing something as harmless as fishing or, indeed, any of the tasks related to such a gentle sport. Unfortunately they do, and not just to anglers who fish out in the wilds where one would expect an emergency to occur.

Some tasks in management are more dangerous than others – for instance, using a chainsaw would be one of them. At all times only *the owner* or the hirer of the chainsaw should use it – and here we assume he has some experience. If you are doing something of that nature you are well advised to have a 'mate' with you, if only as a means of calling for assistance should things go wrong. And at all times it is a matter of common sense to let someone know where you are going to fish and when you should be back, especially where you are going to spend a day wading out in the wilds or when flood water is in a river.

ACID WATER

Fairly intensive research into this relatively new problem has failed to produce a definite cause, let alone a cure. Fish deaths from areas of Scotland and Wales – even highly remote areas – have been sudden, heavy and may herald on-going fishless levels as stocks cease to become self-generating.

Fall-out from factory and power stations chimneys which emit gaseous oxides of nitrogen and sulphur into the air may be the cause, being carried for hundreds of miles before falling and being absorbed into the watershed. Rain falling over large tracts of forestry which drains through largely acidic soil before shedding into rivers has been named as another possible cause of this worrying death problem.

Whether fish become immune, developing an immunity that will enable them to 'breathe' and survive in acid water remains yet to be seen. Certainly we hope that such pollution will not achieve the proportions it has in Scandinavian waters.

80. Abstraction. Here there is no loss of water – it is being pumped up from a small pond at the foot of the hill, and much of it will drain back down again. But on a bigger scale water is tapped from boreholes that will, in time, bring an appreciable reduction of water available to feed local rivers and streams.

ADDERS

Naturalists constantly tell us how charming snakes are and they are far more frightened of us than we are, or should be, of them. It is a statement that we need to challenge – Ken, for instance, holds the world's 20 yard sprint record with full fishing tackle after nearly sitting on a snake, species unknown. And Barrie trod on an adder whilst working a Lake District trout stream. His wellington boots stopped him being bitten.

In theory the adder prefers dry ground whilst the harmless grass snake prefers moisture. In practice no one has bothered to tell the snakes, so it's not necessarily so. Trouble is, the lone angler on a small water knows his business and moves quietly. This renders him liable to sudden contact with any of the reptile tribe.

Our approach is to make a few enquiries and to keep a weather eye open on a new water. Experience has shown that there are definite 'snake' areas where one stands a fair chance of an eyeball to eyeball confrontation. Once this is known it is a simple matter to keep a weather eyeball open and make plenty of undergrowth disturbance as you go along – without disturbing the fish by stamping.

BEES AND WASPS

Singly or in two's or three's, no problem, but in larger numbers, or swarms, they are a worry. Ken has one water where a delightul pool is in the flightline of a wild bees' nest set in a hollow tree on the opposite bank, and that produces some hairy moments. Wasps can be worse – every year there are reports of people badly stung or even killed through disturbing a wasps' nest.

So-called 'wild' bees can often be re-housed and eventually rehabilitated by a bee-keeper, so help from that quarter is often extremely welcome, especially in early season. Wasps' nests can be tackled with a proprietary killer at dusk, when they are all aboard the nest so to speak. Don't undertake this job unless you really know what you are doing – advice and help can often be obtained from the Local Council's Pest Officer.

BOGS

Many people refer to any marshy area of ground as a bog. There is a vast difference! Our experience is that a marsh area is where you sink up to the tops of your waders, then stop. Bogs are where tractors disappear to cab roof level – and carry on sinking.

Marsh areas beside a water can be drained by trenching and lining with ballast, drainage pipes or even branches of hawthorn and other

81. Here, in Cumbria in the drainage basin of the River Lune, numerous channels have been cut as part of a hill drainage scheme.

brushwood laid lengthwise and buried. Water that is drained will obviously lead away into the stream or lake nearby, which means that arrangement must be made for a silt trap.

Bogs are a non-starter. Mark their area, note bearings around them so that you don't, in the dark, walk into one and take care to warn others who may use the water of their existence. In the region of mountain trout streams they are not uncommon even in the present era of over-savage upland drainage.

BULLS

Small water fishermen usually travel light so the temptation to risk a few casts where a bull is in occupation often cannot be resisted; with little tackle you feel that you can outrun the opposition. Sure, the bull may be 100 per cent reliable, but remember that every bull, like a dog, has its day. Michael, Ken's farming friend from Chapter 7, sums it up with a notice on his farm gate:

'If you can cross this yard in 10 seconds – don't.

The bull does it in 9.'

CADDIS LARVAE

A reminder of our instruction in the insect section of Chapter 2. If you transfer caddis larvae or use it for bait, do it in the medium of wet water weed or wet moss. Never use water – otherwise the larvae will asphyxiate.

CONCRETE

Can cause poisoning if applied in a small water and allowed to leach whilst it is wet. Wait till summer if you want to use cement, when water levels are low and you are unlikely to unintentionally pollute.

COYPU

Large, unpleasant looking rat-like mammal that is a pest in certain parts of the country. Does no great harm other than to riddle the banks with holes that ultimately cause them to break away and collapse. The local Water Authority will – or should – assist in trapping, which is the only cure for this problem.

CORMORANTS

We rarely blow our top with any form of wildlife, but this vulture-like sea bird has, in recent years, taken up residence on many of our still and running fresh waters, often at some distance inland from the coast. They eat their own weight in fish every day – and often more than that.

They are at the time of writing protected, although deserving subjects, for an ounce of No.5 shot delivered with the assistance of a 12 bore shotgun. One lake near Cambridge currently has thirty of the brutes in permanent residence, on a small island in the middle of the lake, totally protected by the local bird-watchers who do not wish to go to the seaside to see them.

EELS

Good sporting fish when hooked on the right tackle and we have had some good days (and nights) fishing for them. Even more important, they are excellent eating and exceptionally high in protein value. Whether, as many suggest, you can assist your sex life through their consumption is a matter we leave to the experts.

Eels do consume an enormous amount of fish spawn and on some waters, more especially those immediately above tidal reaches or in upland trout streams, they can be a pest. Most Water Authorities issue trapping licences for a small fee and these traps are not the most difficult of things to operate. Youngs of Misterton, Nr. Crewkerne, Somerset, sell them and also a booklet giving instructions on their setting. The cost of the traps can easily be recouped through the sale of eels, for which there is always an enormous demand – as your local fishmonger will confirm.

FISH TRAPS

Permission to set them is necessary from the local Water Authority who are not over-keen, at least in our experience, on their use. Youngs of Misterton (see Eel entry) can usually supply them or they can be made from galvanised netting and heavy gauge wire, much on the same principle as a minnow trap but on a larger scale.

Frankly, we find little use for them; if you want to get rid of fish then netting is quicker and more certain, except in snag-ridden waters. Traps need regular visiting (night and morning) and this, over a period, is more time consuming than a day's effort with the net, but remembering some recent netting days, perhaps that's overstating things a mite.

FLY BOARDS

An essential on any water, game or coarse, still or running. Made from timber a foot or so long, painted green on top and anchored in the water a little offshore, they are bases for flies to lay their eggs on the underside and provide food for fish.

FROGS AND TOADS

The spawn of both, together with tad and tadpoles are fish food, likewise small frogs – a traditional bait for large chub – and large frogs, ditto for pike. This makes them worth while tolerating, if not actually encouraging. But remembering that both frogs and toads are sure-fire food for snakes ... see entry under Adders! Seriously, the more frogs and toads the better. Herons eat them, which distracts them from cyprinids.

GROUNDBAITING

Something which we use in quantity without a thought on big waters. Perhaps this is why, on small waters, there is a tendency on the part of many anglers to overdo the aid. Remember that groundbait needs to be carried, disturbs fish on being spread and can fill the fish with food very quickly. The rule is not too much, not too often and well in advance, as far as pre-baiting goes. During fishing, a little and often may be necessary.

HERONS

A negative when balanced on the fishery management scales. They consume an enormous number of eels (themselves fish spawn eaters) which may outweigh the fish that herons take. They can, however, be a pest in trout rearing pens and on shallow sections of small trout waters in the early season.

Fortunately they are among the most wary of all birds and a scarecrow, moved around every few days, will keep them at bay. Yet another cure is the regular report from a bird scarer – providing you yourself – and the fish – can stand the disturbance.

INTRODUCING WILDLIFE

Rather a loose heading, but one designed to make all amateur and some professional fishery managers and improvers think very carefully before wading in (rather an appropriate expression) to introduce new life on, or in, a water.

There is a great temptation to make an introduction just for the sake of trying something new, perhaps shoving a new fish of a new species in here, playing around with a strange mollusc there – one could go on forever. If it is not necessary, if it is not really going to achieve anything concrete and improve the water then don't, for God's sake, do it. In the past we have been saddled with the little owl, grey squirrel, and through accident, at least, mink and the coypu – not to

82. The eel trap. Bait (fish, meat, which need not be rancid) is placed into the trap through the open door shown on the side. The trap is then lowered and left overnight around a structure of some sort – bridge, drain supports – or out in between weed runs. Late August, September and October are the best months, the biggest hauls naturally coming from tidal stretches or those immediately above brackish water.

187

mention Canadian water weed, zander and grass carp. They are all mixed blessings that might not, with careful thought, have been necessary, to say the least.

KINGFISHERS

Where would we be without them. Bright flashes of colour to brighten dull moments. They can, like the heron, prove a pest where intensive rearing is practised and will whip small trout out of a stock pen like lightning. This can be cured and the birds persuaded to move on with the aid of a plastic decoy owl, which is sold by most shooting stores. Watch where the kingfisher perches – he will have one favourite spot above the water to which he will return – and install the owl close to it. The cure is nearly 100 per cent sure.

LANDING NETS

Don't scale the size down just because you are fishing on a small water. Fish grow big on small waters as well as on large ones and a big net is often required, especially where pike are concerned. Round frames get less tangled in bankside growth; triangular frames trap better in shallow water.

MINK

Nasty little things, vicious when handled and demanding on fish life. The Water Authority will usually supply traps, but the best cure is lead poisoning implanted by means of a cartridge. Never attempt to handle mink alive, and remember they can feign death.

NEWTS

An amphibian to which the angler pays scant attention. These delightful little things are fish food however, but we hasten to add that we could never mount one on a hook, but in America they use plastic salamanders (somewhat similar in outline) as a bait as well as live ones, and it might not be so long before they appear on this side of the Atlantic, hailed as a 'super new attractor'.

POACHERS

Throughout this book we have quoted our experiences with this growing army of angling freebooters. A tangle with them can completely spoil a day out and may even result in violence against an innocent party.

Our attitude to the problem is that what cannot be seen cannot be acted on, and we try to complete each job on a fishery with as little disturbance as possible, resisting all temptation to put shine and polish on what we do.

Deliberate traps – underwater varieties that block a swim, not pits with sharpened stakes set in the bottom – are also a help and discourage repeated visits. Through sheer frustration we have, in the past, ensured that a few poachers with cars have had a long walk home; rather a silly attitude on our part perhaps as it puts one within the grasp of the law and reduces oneself to the level of the poacher himself.

Our present line, after polite reasoning has failed, is a stream of bricks and stones from a safe distance around the water in front of the angler. It spoils his chances for several hours, and is a relatively safe ploy as he will be loath to get up, leave his tackle unprotected and give chase. No good getting old unless you get crafty!

POLLUTION

Hands up all those who know the telephone number of the nearest Pollution Officer of the Water Authority! You don't? Then write it out 100 times and make sure you use it when you suspect, but are not necessarily sure, of an outbreak.

RUBBISH

Rubbish attracts rubbish, so if your fishery is near an urban area then make sure you remove or destroy that which appears, even if you didn't put it there in the first place. Rubbish from the fishery, weed from cutting expeditions, old branches etc. should be burned completely or buried well away from the water's edge. It's an even bet that if you don't do this someone will throw the stuff back into the water.

SILAGE PITS

An unusual heading in a fishing book but a necessary reminder to all with a water close to farm buildings. Silage pits are washed out yearly and the water, which has to drain somewhere, is toxic to fish.

If you have a fishery where this could occur make a point of introducing yourself to the farmers and let them know that you are working on, and trying to improve a fishery close by. A jolt to the memory about the results of pit washing, during the conversation, could save some nasty problems.

STOCKING

Don't attempt a Do-It-Yourself effort without necessary permission from your local Water Authority. By far the easiest way of going about the business is to get a quote from a recognised fish farm for the fish you require. They will supply certified disease-free fish from stock and will help – or even take over – negotiations with the Authority for the stocking.

Remember also that the Authority will net and re-stock; often they have fish to spare and are looking for somewhere to put them. Nothing venture, nothing gain and a call with an explanation of your wants costs nothing.

WADING

A short caution under the heading of Accidents has already been made. Remember never to attempt to wade straight up or downstream. Work diagonally and don't be afraid to use a wading staff if the current is strong. Even young men use them.

WEATHER

Something we anglers take for granted. It's always there, damn it! It makes a great impression on a fishery, however, and it is worthwhile recording conditions in a diary form at any time you visit the water. We wish we had been more astute at doing it during the early days of some waters we use today.

WEED RAFTS

Often an unwanted present from those upstream of you. Quite apart from the fact that they may ruin a day's fishing, they are also an enormous source of fish food going to waste. Most will pass down stream but those that are trapped on your water need to be removed. Do it in stages if possible, giving some of the insect life on board a chance to leave and take up residence. After 24 hours the rafts pass under the heading of Rubbish and should be dealt with accordingly.

FURTHER READING

The following references are useful not only in dealing in part at least with aspects of small water management and fishing but in enabling the reader to delve widely into a somewhat scattered literature. The list is far from complete, but these books and papers themselves contain references in a minority of cases. The publications of the Institute of Fishery Management should also be perused and the serious manager would do well to join that organisation, at least in its local branch, as well as the Freshwater Biological Association.

BIBLIOGRAPHY

Atkinson, B., 1961 *Angling from the Fishes' point of View* Herbert Jenkins
Berhrendt, A., 1977 *The Management of Angling Waters* André Deutsch
Bracken, J. J. & Kennedy, M. P., 1967 *A key to the identification of the eggs and young stages of coarse fish in Irish waters* Scient. Proc. R. Dubl. Soc., 2, 99-108
Clapham, A. R., 1947 *Trout fishing on Hill Streams* Oliver & Boyd
Clapham, A. R., Tutin, T. G. & Warburg, E. F., 1952 *Flora of the British Isles* Cambridge University Press
Clegg, J., 1952 *Freshwater life on the British Isles* Frederick Warne
Clegg, J., 1956 *The Observer's Book of Pond Life* Frederick Warne
Dawson, K., 1928 *Salmon and Trout in Moorland Streams* Herbert Jenkins
Drummond Sedgwick, S., 1976 *Trout Farming Handbook* Seeley, Service
Ellis, E. A., 1946, 1947 *Freshwater Bivalves* Linn. Soc. Synopsis of the British Fauna nos. 4 & 5
Evans, J., 1972 *Small-river Fly Fishing for Trout and Grayling* A. & C. Black
Fitter, R. & Fitter, A., 1974 *The Wild Flowers of Britain and Northern Europe* Collins
Forbes, D. C., 1966 *Small Stream Fishing* Newnes

Forbes, D. C., 1977 *Rough River and Small Stream Fishing* Cassell
Frost, W. E. & Brown, M. E., 1970 *The Trout* Collins
Garrow-Green, G *Trout Fishing in Brooks* Routledge
Gregory, M., 1976 *Angling and the Law* Charles Knight
Harris, J. R., 1956 *An Angler's Entomology* Collins
Hutchinson, G. E., 1975 *A Treatise on Limnology* Wiley
Ingham, M. & Walker, R., 1964 *Drop Me a Line* Macgibbon & Kee
Inglis Hall, J. *How to Fish a Highland Stream* Putnam
Jacques, D., 1965 *Fisherman's Fly* A. & C. Black
Knowles, F. G. W., 1953 *Freshwater and Saltwater Aquaria* George Harrap
Lawrie, W., 1947 *The Book of the Rough Stream Nymph* Oliver & Boyd
Macan, T. T., 1949 *A Key to British Fresh and Brackish Water Gastropods* Fresh. Biol. Ass.: Sci. publ. no.13
Macan, T. T. & Worthington, E. G., 1974 *Life in Lakes and Rivers* Collins
Maclean, N., 1980 *Trout and Grayling* A. & C. Black
Mellanby, H., 1938 *Animal Life in Freshwater* Methuen
Newdick, J., 1979 *The Complete Freshwater Fishes of the British Isles* A. & C. Black
Phillips, R., 1977 *Wild Flowers of Britain* Pan Books
Platts, W. C., 1977 *Trout Streams, their Management and Improvement* Harmsworth
Price, S. D., 1976 *Rough Stream Trout Flies* A. & C. Black
Reid, D. H., 1944 *Key to the Families of British Gammaridea* Linn. Soc. Synopsis of British Fauna, no.3
Righyni, R. V., 1968 *Grayling* Macdonald
Scott, J., 1969 *Sea Trout Fishing* Seeley, Service
Skytte Christiansen, M., 1979 *Grasses, Sedges and Rushes in colour* Blandford Press
Spencer, S., 1935 *Clear Water Trout Fishing with Worm* Witherby
Spencer, S., 1968 *Salmon and Sea Trout in Wild Places* Witherby
Tudor, F. E., 1955 *Trout in Troubled Water* Herbert Jenkins
Van Duijn, C., 1967 *Diseases of Fishes* Iliffe
Wells, L., 1937 *Garden Ponds, Fish & Fountains* Frederick Warne
Wheeler, A., 1969 *The Fishes of the British Isles and North West Europe* Macmillan
Wright, D. M., 1973 *The Fly Fisher's Plants* David & Charles

INDEX

193

196